# *Talleur's Basic Fly Tying*

***Books by Dick Talleur***

•

Fly Fishing for Trout

Mastering the Art of Fly Tying

The Fly Tyer's Primer

The Versatile Fly Tyer

Talleur's Dry-Fly Handbook

Modern Fly-Tying Materials

Talleur's Basic Fly Tying

# Talleur's Basic
# FLY TYING

*Dick Talleur*

The Lyons Press

Printed in Canada

10 9 8 7 6 5 4 3 2

***Design by M.R.P. Design***
***Typesetting & Composition by CompuDesign***

Library of Congress Cataloging-in-Publication Data

Talleur, Richard W.
[Basic fly tying]
Talleur's basic fly tying / Dick Talleur.
p. cm.
Rev. ed. of: Fly-tyer's primer. c1986.
Includes index.
ISBN 1-55821-519-0
1. Fly tying. I. Talleur, Richard W. Fly-tyer's primer.
II. Title.
SH451.T286 1996
688. 79'12—dc20

96-12521
CIP

# *Contents*

Introduction *vii*

**1** Tools and Instruments *1*

**2** Getting Started *17*

**3** Wet Flies *33*

**4** Nymphs *51*

**5** Streamers *63*

**6** Dry Flies *91*

**7** Putting It All to Work *125*

**Appendix** Hook Substitution Chart *149*

Index *151*

# Introduction

Welcome to fly tying. First, I want to assure you that you will be successful at it. Fly tying, given the right materials and competent instruction, isn't at all difficult. Golf and tennis are difficult, and to become really good at those games presupposes a generous helping of talent. Not so with fly tying; anyone with adequate vision, corrected or not, can become proficient.

The reasons for producing this book, which replaces *The Fly Tyer's Primer,* are simple: Far better graphics are now available, and economically feasible, within the world of publishing. Thus, we're able to produce, at an affordable price, a full-color book. Also, I now have 10 more years of teaching under my belt and have fine-tuned and enhanced that process.

The fly patterns that follow are all very effective ones. I don't think it would be either instructive or enjoyable for you to sit and tie flies that have only academic value. However, I also want to assure you that they were chosen with teaching in mind. They require the fundamental skills in technique, material selection, and materials-handling that are at the very heart of fly tying.

Except for the first section, which addresses tools and instruments, I've avoided lengthy, abstract narratives. Instead, I've incorporated the explanations of things into the pattern lessons themselves, so that you can readily see the relationships between the components and how they're used. I think it's a lot more useful to put a hook in a vise and tie some stuff to it than it is to read a long dissertation about the history of hook-making.

One of the most important lessons I hope to teach is how to deal with the architecture of fly tying. By this I mean cohesion: being able to attach fly components in an integrated manner, so that everything works out and the overall effect is what you desired. This is very much related to transitions, which implies working through a pattern in an intelligently progressive manner.

Fly tying is something like playing billiards; what you do now affects what you'll be faced with down the line. The idea is that each step accommodates those that follow. This is why you'll notice that the book is quite explicit about quantities of materials, positioning of thread, and such. This is the essence of the art and science of fly tying: setting up what comes next. Failure to learn basic skills causes real problems. The idea is to solve such problems by avoiding them altogether. That's when fly tying becomes fun.

A word about the photographs. Any of you who've ever worked in macrophotography know that a 1:1 macro lens is about as unforgiving as a reticent mother-in-law. The camera sees things that the eye of the fly tyer can't, little filaments of a frayed piece of floss, stuff like that. Also, small details that are quite inconspicuous to the naked eye become magnified through the macro lens and thus visibly distorted.

I mention all this not by way of trying to excuse any less-than-perfect tying I may have done in the photographed sequences, but rather to inform you that your fly might not—and need not—look exactly like the one in the book in every excruciating detail. I don't want to cause a phenomenon like the one that occurred when the Japanese began to record American jazz. They were so meticulous, and so attentive to detail, that they even copied the mistakes!

A couple of special acknowledgments: First, to Tim Savard, for his brilliant photography in the full-page layouts in chapters 1 and 7. Second, to Paige Wilder, for her excellent drawings, which enhance the book.

On that note, I'll simply add that I very much hope you'll find this book to be both instructive and fun. Enjoy.

—Dick Talleur
1996

◀ ONE ▶

# *Tools and Instruments*

The beginning fly tyer will need to acquire the tools of the trade, so to speak. I'll treat this subject thusly: first, a listing of and descriptive information about those tools that are absolutely essential; then, a supplemental listing and description of some others that, in my experience, have proven to be of value.

## Fly-Tying Vises

Today, there are two distinct types of vises: stationary and rotating, or rotary. However, within the stationary category are vises with the ability to revolve. It's important to be aware that these aren't the same as rotating vises. A functionally rotating vise is one that facilitates *full and free 360-degree rotation, with the hook on the same axis as the center line of the jaws and shaft of the vise, and is designed so as to allow the tyer access to the hook from the rear.*

While of benefit to the experienced tyer for certain applications, true rotating vises require a more sophisticated technique and, in my opinion, are not for the beginner. If at some point you want to know more about them, I refer you to another book of mine, *Modern Fly-Tying Materials,* offered by the same house that publishes this volume. Here, let's look at the more traditional type of vise, the stationary model.

### *Mounting Mechanisms*

The choice of C-clamp or pedestal mount is up to you. Here are the main considerations:

- The C-clamp allows adjustment of the vise's height to match up with the chair used by the tyer and his or her physical characteristics.
- The C-clamp is lighter in weight, a consideration for the traveling angler.

- The pedestal obviates worry about thickness, shape, and "overhang" of the desk or table; it will sit on any flat surface.
- The pedestal doesn't interfere with lap drawers.
- With the pedestal, the vise is situated over the tying table, and dropped items usually land there. With the C-clamp, the vise is mounted in front of the tying table, and dropped items usually wind up on the floor or in a waste receptacle.
- The C-clamp may be of benefit to the novice tyer, who doesn't always apply thread tension directly downward.

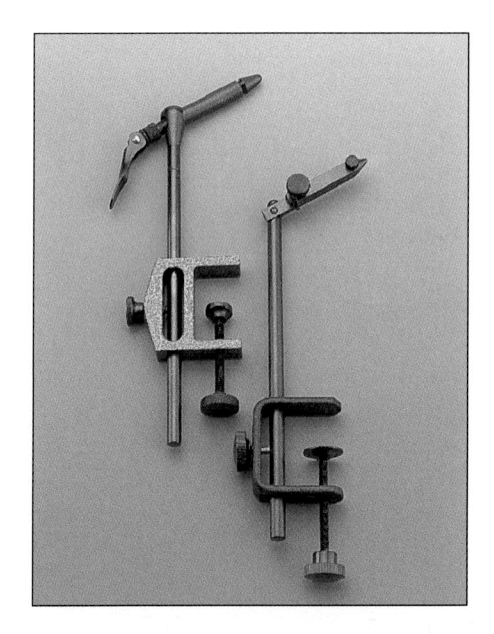

***Two popular, serviceable, moderately priced vises. On the left, the Thompson Pro; on the right, the Griffin 1-A. Both have C-clamp mounts.*** Photograph by Tim Savard.

A qualifying statement about the first consideration. While pedestal-mounted vises aren't height-adjustable per se, it is possible to correct your position relative to the vise as follows: If the vise is too low, put something under the base, such as a book; if the vise is too high, put a thin cushion on your chair seat, thus elevating yourself. I might mention that an office chair with a height-adjustment feature is ideal for fly tying.

Many vises offer conversion kits that enable you to switch between the two types of mount. This process is simple enough, but somewhat costly, as it also requires you to change mounting rods, the C-clamp rod being considerably longer than the pedestal rod.

### *Features*

First, a list of the basic features that every vise should have:

- Adjustment for hook-wire diameter (size of hook).
- In C-clamp models, sufficient width to accommodate most tables.

- Collet or hook-holding mechanism angle adjustment.

Secondary but still desirable features:

- Revolving capability.
- Interchange between C-clamp and pedestal.
- Interchangeable jaws.

*Collet angle adjustment* means the ability to make the hook-holding part of the vise more or less horizontal. Note that I put it in Group 1. I might get some arguments over this, but in my experience the ability to make this adjustment is important, as it enhances access from the rear to various sizes of hooks. Smaller flies are becoming more popular all the time, and tying them is not at all difficult if you can get at the hook with your fingers. There's an interaction between this feature and revolving capability, which I'll discuss further subsequently.

## Jaw Closure Mechanisms

Vise manufacturers employ various designs for closing the jaws. The most common is the cam-lever/draw-bar type, in which the downward operation of a lever, interacting with a simple wiper cam, causes the jaws to be drawn rearward into a tube, or *collet,* thus closing them. In some lower-priced vises, a screw wheel is substituted for the lever, which helps keep costs down. Rotating this wheel creates the same effect as pushing down a lever. It's a little slower, but it works just about as well.

Another design also utilizes the screw principle. It consists of a round knob or wheel located near the front of the jaw mechanism that, when turned, causes the jaws to close. However, it's quite different from the mechanism I referred to earlier in that there's no cam, collet, or draw bar. Instead, there's a simple lever, which behaves something like a seesaw. When one end goes one way, the other end goes in the opposite direction. In this case, it's an asymmetrical seesaw, but it employs the basic principle of leverage. The end that effects the jaw closure is much longer than the jaw itself, which enables you to apply sufficient force to hold a hook effectively.

This mechanism requires the use of a second, smaller screw knob, the purpose of which is to allow you to change the *dimension,* or distance between the two sides of the jaw at the point of axis. This is an important part of the design, as it facilitates adjusting for various wire diameters. You must become familiar with this adjustment feature if your vise is to accommodate a wide variety of hook sizes.

This operation may sound a bit tedious but, in practice, it's pretty simple once you become accustomed to it. This design is employed on several low-priced vises and also on some sophisticated ones.

The Regal vise produced by the Regal Engineering Company employs an internal spring mechanism for closing the jaws, which it does automatically. In other words, you apply pressure when opening the jaws rather than when closing them. Then you relax your grip on the lever, and the spring does the rest.

The standard jaws have a curved groove inside to hold hooks with absolute security. There's a degree of compromise here in that the shape fits certain hook bends better than others. However, this can be rectified by adjusting the angle of the vise proper, so that the hook shank lies on a flatter plane. For holding larger hooks, I must give this feature an excellent rating.

However, you can't use the groove on medium-to-small hooks, and certainly not on dry-fly hooks. But you don't need it: Simply mount those hooks up front, ahead of the groove.

The Regal comes in both pedestal and C-clamp models and has both angle-adjustment and revolving capability.

***Two more vises, also popular. On the left, the Regal RPS-7; on the right, the Voyager, from the Kennebec River Fly & Tackle Company. Both have pedestal bases.*** Photograph by Tim Savard.

## *Jaw Options*

Jaw design is, of course, of great importance. There's more to this than simply accommodating hook size, which in this context essentially means wire diameter. There's also the matter of shape, which affects both hook-holding and hook-access capabilities. Shape interacts with the important angle-adjustment feature found on better-quality vises.

Some vises offer interchangeable jaws. One model in particular, the HMH (Hunter's Multi-Head) produced and distributed by the Kennebec River Fly & Tackle Company, is outstanding in that it offers the convenience of screw-in/screw-out jaws. While not necessary for getting-started tyers, this is a most valuable feature for the more advanced, whose tying has become more diversified, so I'll cover it briefly.

The HMH comes with a more or less standard jaw that will accommodate a wide range of hook sizes, including all that appear in this book. At additional cost, you can purchase additional jaws that are at the opposite ends of the HMH spectrum: very heavy ones for huge hooks and very delicate ones that accommodate tiny hooks. The delicate jaws are a joy when you're tying on size 20 or smaller. If at some point you should acquire these, please resist the temptation to leave them in place when tying larger sizes, as jaw damage will almost surely result.

Some manufacturers offer jaws and collets that are matte-finished rather than chromed. Personally, I prefer this, as I find it restful on the eyes.

### *Adjustments*

For beginning tyers, the most important adjustment—the one you most need to understand—is the one that facilitates altering the gap of the jaws to allow for finer- or heavier-wire hooks. To my knowledge, all vises on the market today have this capability. The mechanism will vary from make to make, and you should familiarize yourself with the instructions for the one you decide to buy.

As I mentioned before, it's beneficial to be able to adjust the angle of the hook-holding part of the vise, especially when working on smaller hooks. Many stationary-type vises have a revolving feature whereby the collet/jaw assembly can be turned 360 degrees, thus allowing you to inspect, and perhaps even work on, the backside of the fly. This is useful for tyers at all levels; however, as I've stated, there's an interaction here.

When you're using larger hooks, you can set the angle to a flatter plane, making the revolving feature more viable. This still doesn't equate to or approximate full rotating operation, but it does allow you access to the far side of the fly, if for nothing more than a visual check and some moderate tweaking. The more severe the setting of the angle adjustment, the more difficult it is to use the revolving feature, because the jaw/collet and the hook shank are on different axes.

Some of these angle adjustments require one or more tools—usually a small screwdriver and hex wrench. The more sophisticated vise models may require several hex wrenches of different sizes, which are, in most cases, included by the manufacturer, along with instructions. Keeping track of these items is a pain, especially when you're traveling. I've always felt that a manufacturer should design and produce a specialized, unified tool set, perhaps along the lines of a multibladed jackknife. So far, I haven't been able to convince anyone.

Getting back to jaw shape, there are shapes that can be considered more or less universal in that they will, with proper adjustment, accommodate a wide range of hook diameters and allow adequate access to fairly small hooks. It is feasible to mount fairly heavy hooks in jaws of this type if you're careful to position the hook farther back in the slot, where the metal is thicker and stronger. Of course, strict attention must be given to jaw adjustment, and you mustn't force anything.

There are vises to fit every budget. The less costly ones are adequate for typical work. Naturally, you can't expect the same precision, ease of adjustment, durability, number of features, product life, or overall performance from, say, a $30 vise that you can from a $200 model, but for a young beginner or occasional tyer, it will do the job. Perhaps the main thing to know before you buy is what sort of hooks you'll be tying with. Verify that the vise you're considering will readily accommodate them.

### *The True Rotary Vise*

While I don't recommend this type of vise for beginning tyers, it does hold a position of prominence in today's market, and I'd be remiss if I didn't make mention of it. Automation of certain fly-tying operations has long been the dream of vise designers and manufacturers, and over the years many have striven to perfect a device that accomplishes this. Let's take a look at a modern rotary vise and see how they've done.

As of this writing, the most popular rotary vise is made by Andy Renzetti. There are several models available with a wide diversity of prices and features.

These vises must be specifically ordered for either left- or right-handed operation. Either C-clamp or pedestal mounting is available. They have a tension adjustment, which controls the amount of resistance the main shaft offers to rotation, and can operate in either the rotating or stationary mode.

The two main advantages of this type of vise are:

1. Full rotation, for increased speed on tedious operations such as wrapping stranded material onto a large hook.
2. The capability of inspecting and working on a fly from any vector while maintaining it in a level position on a true axis.

Now for the other side of the coin:

1. Hook mounting and adjustment are a bit slower than with stationary cam-lever vises.
2. Because rotary operation necessitates keeping everything on a single center line, the superstructure of the vise somewhat inhibits rear access to the hook.
3. Operation in the rotary mode requires that the thread bobbin be suspended in front of the jaws in the bobbin rest. This mandates securing the thread by either a couple of half-hitches or a three-turn whip-finish knot.

I must state—and this is my personal view—that I am bothered by the last two items. I like to be able to get at a hook from the rear without having to alter my hand position or think about dodging anything. I also find having to fool with the bobbin rest cumbersome, and I particularly dislike having to secure the thread by tying a knot. In many instances, after making a double-wrapped body with floss, tinsel, or whatever, I want to *back off* the thread—that is, to unwrap the wraps I tied on with, so as to reduce bulk and keep everything smooth. With the rotary, not only can't I do this, but I also have an additional knot to deal with.

Nevertheless, I must in fairness state that the Renzetti line of products is, from a standpoint of quality and workmanship, excellent. There is another rotary vise on the market, the Nor-Vise, that I haven't used personally but have heard good reports about.

### *Care and Maintenance*

This will be a short treatise, because fly-tying vises don't need much in the way of care. In fact, your main task is simply to follow previous instructions regarding adjustment. The one thing that can hurt or ruin a vise is forcing the jaws to close on a hook too big for its setting. This stresses several components, particularly the jaws. Not only will you impair the instrument's ability to perform its function, but you may also invalidate the warrantee.

With conventional vises, adjustment is proper when full closure is obtained with the cam lever just past the vertical: in other words, somewhere between the five o'clock and six o'clock positions. Check the instructions that come with your vise for further details.

With vises that use the two-screw mechanism, the procedure is to set the adjustment with the front screw, then effect closure with the rear one. Don't try to cheat. The Regal needs no adjustment. Vises that employ other closure mechanisms come with instructions. Heed them.

## Tools

As you'll soon learn, there's a vast array of fly-tying gadgets on the market today. Here, I'll discuss only the essential ones, plus a few I feel are very helpful. I advise discretion regarding the more fringy.

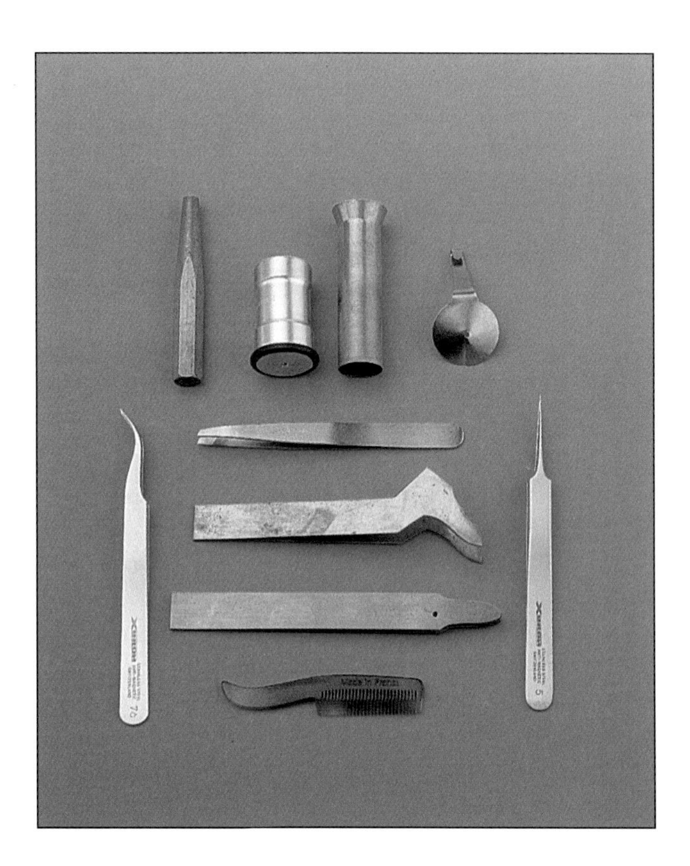

***An array of tools. Starting at the top and working clockwise: the Matarelli whip-finish tool; a pin vise set up as a bodkin; two English hackle pliers, one with plain jaws, the other with heat-shrink tubing on one side; the large Dr. Slick scissors; iris scissors; Matarelli bobbin; bobbin threader/ reamer; Griffin bobbin; Umpqua ceramic-tube bobbin.***

Photograph by Tim Savard.

### *Scissors*

There's a great variety of scissor functions and prices from which to choose. If you're a typical trout- and salmon-fly tyer, you'll want your number one scissors to be fine-tipped, very sharp, and of excellent quality. Later, you may want to buy another pair for rough work to protect and preserve the better ones.

Don't buy cheap scissors; they're a bad investment and a functional disaster. They won't hold an edge and will disintegrate with sharpening. They're also vulnerable to being sprung under only moderate stress.

Comfortable-fitting finger loops are an important factor. Comfort is a matter of personal preference. Some people like large loops and a loose fit; others, myself included, prefer a tighter fit for better control.

When tying, I keep my scissors in my right hand virtually all the time. However, I don't want them to slide all the way down to the base of my fingers; I want them seated just below the first knuckle of my fourth finger and about at the bottom of my thumbnail. My hands are neither large nor small; I'd say they're about average. Thus, I don't require very large finger loops.

Guys with large hands may need larger loops. There are a number of scissors on the market that have oversized finger loops, as well as several brands that have adjustable finger loops.

Fly-tying scissors come with either straight or curved tips. For beginners, I suggest straight tips, but the choice is yours. Personally, I've come to prefer the curved type for most work, especially when frequent discrete trimming and shaping of materials is required.

Scissors are available that have fine serrations along one blade. I absolutely love this. It causes the blades to grip the materials and cut them with dispatch, rather than sliding along them. This feature is particularly effective when you're trimming hair, as on a Muddler Minnow head or a hair bass bug.

As I mentioned previously, the ideal situation is to have at least one pair of heavy-duty scissors at hand, for cutting hair, metallic materials, synthetics, or anything that might take a toll on your precious delicate scissors. I'd guess that more scissors are ruined by being splayed than by being dulled, and trying to cut materials that are too bulky will do just that. I realize that switching scissors is somewhat of an inconvenience, especially if you tie with scissors in hand, as I do; no one likes the interruption of continuity that changing scissors entails. But please try to discipline yourself; your scissors will thank you. Failing this, if you must cut a piece of heavy oval tinsel or the like with your class A scissors, at least do so well down into the blades; stay away from the tips.

### *Bobbin*

This is such a basic and simple tool that little need be said about it. There are only three essential concerns. The first is that you use the tool properly. If you're a beginning tyer, you should develop your skills in handling the bobbin and avoid falling into bad habits, such as handling the thread with your fingers as it feeds out the tube. This makes wrapping more difficult, in addition to possibly abrading the thread.

The second essential is proper adjustment. The bobbin should feed out thread with moderate resistance, but should not be so loose that it feeds the thread by virtue of its own weight. New bobbins invariably need to be loosened up. Beginners may want to watch an experienced hand do this the first time, to get the hang of it. You simply and gently bend out the limbs, one at a time, until you achieve the desired setting. If the limbs rub against the edges of the spool, change the angle of the bend a little to correct that. Now you know how to adjust a bobbin.

The third concern is that the tool be free of defects, which means that the mouth of the tube must be smooth, so that it won't cut the thread. I encounter a lot of this in my beginners classes. If you should notice that your thread frays a lot and breaks easily, try a few little experiments, such as:

- Put another spool of thread in the suspected bobbin, do some wrapping, and see if the problem continues. If not, you probably have a bad spool of thread.
- If the second spool also frays or breaks easily, try them both in another bobbin, preferably one that you've been tying with problem-free. If this solves the problem, you know it's the bobbin.
- If you continue to have problems, take your threads and bobbins to an experienced tyer or to the staff of a local fly shop; have them go through the same experiments. Also, have the tension adjustments checked, as well as your technique; you might be tying a bit heavy-handedly.

The basic bobbin employs a metal thread tube. Other models use a ceramic tube, and still others feature a little ceramic doughnut at the mouth of the tube. The metal ones are the most prone to having a burr or sharp edge within the tube, but now and then this does occur in a ceramic one; I've had it happen to me. No problem; just return the tool.

Bobbins are available with long tubes. I've never found the need for one, but then I don't tie every kind of fly in the whole world. Maybe they provide advantages in bass-bug or very large streamer tying.

### *Hackle Pliers*

Usually, I demur at mandating specific models or types of equipment in fly tying. My preference is to describe the positive and, if any, negative attributes of the various types and let you make up your own mind. In the case of hackle pliers, however, I'm going to make a specific recommendation; but first, some criteria.

The adage that form follows function definitely applies here. Let's consider the two basic tasks hackle pliers must perform:

1. They must grip a feather securely enough that it doesn't slip out, but not so hard that they crush or break the quill.
2. By virtue of their weight, they must maintain enough tension on a wrapped hackle that it doesn't loosen, so that you can let go of the pliers and do something else.

That's it! And still I find that there are pliers out there that don't perform these two simple chores effectively.

Fortunately, the time-tested *English hackle pliers* are still with us. Find out who sells them, and buy them. Get the full-sized kind, not the mini version, which are too small and too light. Tiny hackles can be wrapped perfectly well with regular-sized hackle pliers; it's the jaws that matter.

I also strongly suggest the following embellishment: Visit your friendly hardware store and buy some narrow-gauge heat-shrink tubing. Disengage the jaws of the pliers and slide the tubing over one jaw. Cut it to length and apply a little heat, so as to shrink it tightly around the metal. Reengage the jaws and you have a dependable tool that will not only perform the two basic functions, but many others as well.

For example:

- Making and spinning a dubbing loop.
- Twisting peacock herl with thread for reinforcement.
- Holding the butt end of a feather while you're folding it or while you're extracting barbs for tailing or hackling.
- Grabbing the end of a broken thread so as to maintain tension while you're rebooting.

That about covers the essentials: scissors, bobbin, hackle pliers. Now for the other tools I recommend.

### *Bobbin Threader/Reamer*

This little tool is used for drawing the tying thread through the tube of the bobbin. You can fashion a makeshift one out of a piece of fine wire or monofilament. There is a model, however,

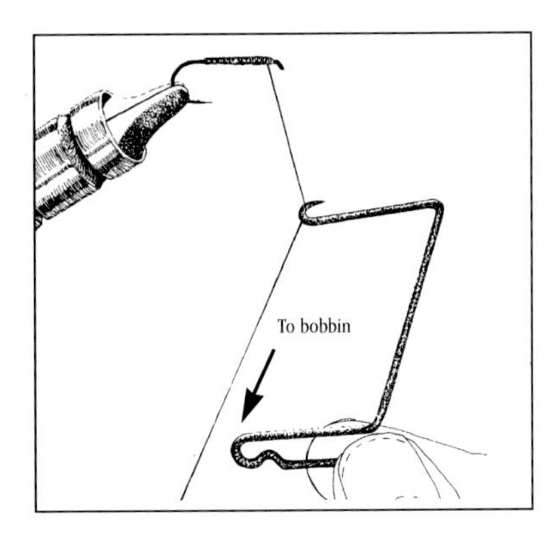


***The starting position—thread "hooked" onto the tool. Be sure to have enough thread run out of the bobbin to allow the tool to work freely. Also, hold onto the round bead attached to the shaft, so that the bobbin cannot rotate within the tube.***

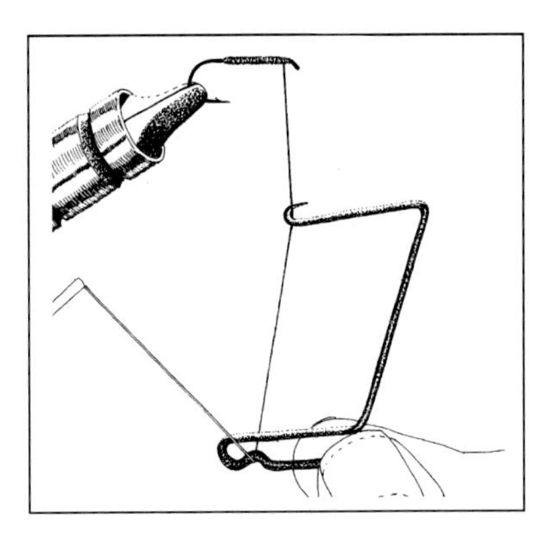

***Thread positioned in the "notch" of the tool.***

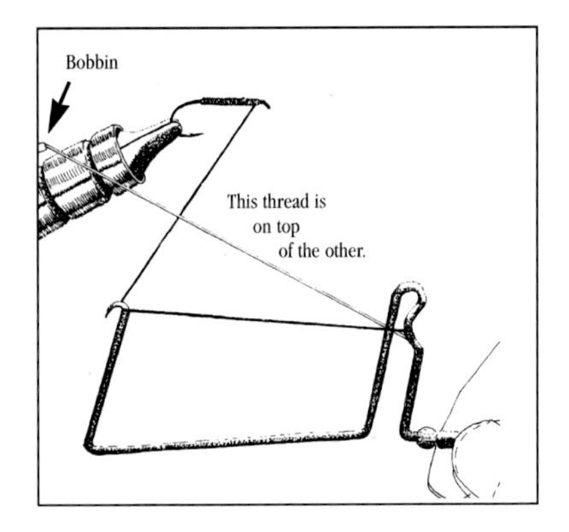


***Release the bead, so that the tool rotates a half-turn. Form a triangle by passing the thread over itself, as shown.***

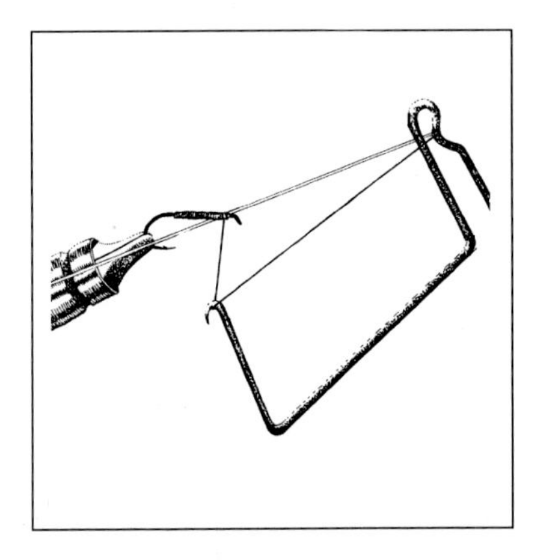

***Position the triangle so that the eye of the hook is between the two threads and in the vortex of the triangle. Note carefully the position of the tool.***

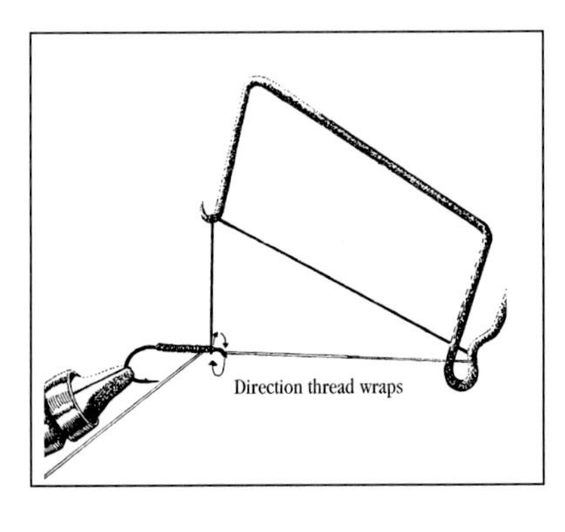




***While holding the tool by the tube, so that free rotation can occur, make five or six neat wraps, as shown. Note that they follow the normal direction of all thread wraps—that is, clockwise.***

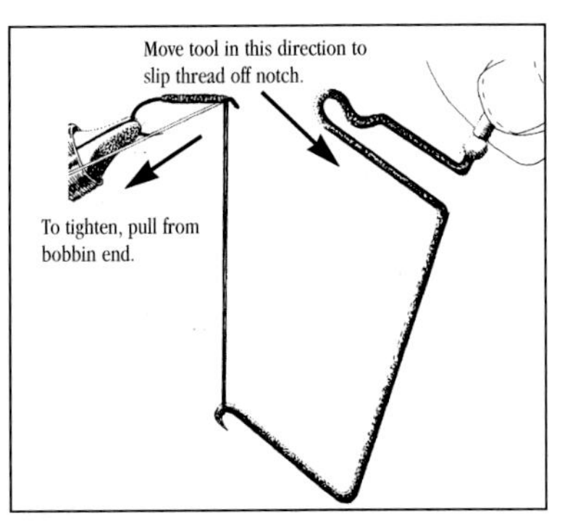


***While maintaining tension on the loop, slide the thread out of the notch. Then finish off the knot, tightening the loop by pulling from the bobbin end of the thread.***

that has a threader on one end and a reamer on the other, with a short piece of bead chain in the middle connecting the two. The reamer is required periodically for cleaning out wax buildup in bobbin tubes. That makes the device well worth buying.

If you use ceramic bobbins, be careful when reaming out the tubes. It's possible to crack or nick them if you're a bit heavy-handed.

### *Whip-Finish Tool*

This device is used for making the finish knot on the head of a fly, something I did by hand for the first 25 years of my tying career. The model I recommend is the Matarelli, because of its unique design. The part of the tool that actually forms the knot rotates within a brass tube. Your first few experiences with this tool may be a bit frustrating, but please believe me: Once you've gotten the hang of it, it's automatic.

As of this writing, the Matarelli is available in two sizes: regular and long-reach. Personally, I use the long-reach for everything. I should point out, however, that the smaller model does just as good a job and is more compact—a feature the traveling tyer always likes.

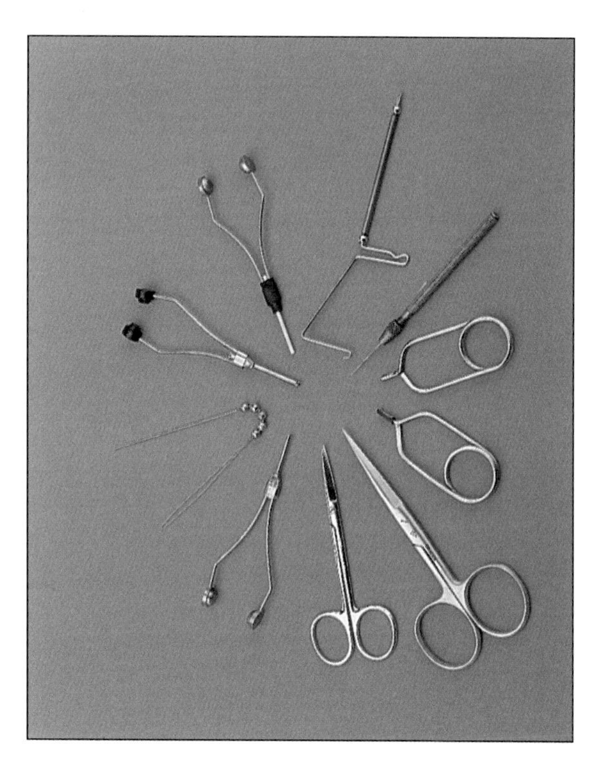

***More tools. Top row: hair-packing tool; Renzetti hair evener (two parts); hackle guard. Center, from top down: Tweezerman flat-nosed model; two different designs of wing burner; small comb. Left and right: two useful models of tweezers.***
Photograph by Tim Savard.

### *Hair Evener*

Often referred to as a *stacker,* the purpose of this tool is to render the tips of a bunch of hair even, which abets the making of nice, neat wings. I highly recommend this tool.

The most important factor in a hair evener is the diameter of the inside of the tube. There are a lot of cute little stackers out there, and they may be okay for working with small bunches of fine, straight hair, but for more sizable amounts—or for hair that's crinkly—you'll want a wide tube. Be careful: Some stacker tubes look a lot wider than they really are, because they're made of heavy material. Remember, it's the *inside diameter* that counts. Renzetti stackers are very good in this regard.

### *Materials Clip*

This handy little device mounts on the vise and is used for holding pieces of stuff, such as tinsel, out of the way until it's time to wrap them into place. The most widely used model today is nothing more than a small spring. The materials are wedged into the slots between the coils.

The problem here is that some materials are of so fine a diameter that they won't stay in the slots. But there is a model—the HMH materials clip, designed by Bill Hunter—that neatly solves this problem. It consists of two springs of different diameters. The smaller one is mounted inside the larger, but it only extends halfway; thus, half the circumference of the device has wider slots, while the other half, by virtue of the double spring, has narrower ones. This materials clip will hold even the finest of tinsels.

### *Hackle Gauge*

This is a device for measuring dry-fly hackle with respect to hook size. Some people have a natural talent for judging proportions; others need a crutch. For the beginning tyer, a hackle gauge is, in my opinion, a most helpful tool. It's rare that a new fly tyer has such visual acuity and sense of proportion that he or she can judge proper hackle sizes right off the bat. Later, as experience sharpens visual skills, the tool can be set aside.

Many of today's hackle gauges have a pin of some sort around which you wrap the hackle to simulate its position on an actual hook. This is good. Several mount on the vise, a convenience.

### *Tweezer*

This is a handy item to have on a tying table. It's quite useful for picking up small objects, such as tiny hooks. It can also be used in certain tying operations.

One in particular has proven to be most useful. The flat-nosed tweezer performs many useful tasks besides simply picking up things and offers the added advantage of durability.

Be careful of bargains. I have some inexpensive tweezers that looked great in the store; as it turned out, however, the metal was so soft that the tips bent with minimal provocation. Considering that these are lifetime tools, it pays to go for quality. The Tweezerman brand is very good.

### *Finger-Smoothing Items*

Burrs, hangnails, and calluses are the bane of tyers the world over. People who do manual work, especially outdoors, often experience serious problems with this. Hand lotions provide only partial relief. Significant burrs that interfere with tying need to be smoothed out more effectively.

A pumice stone will do the job quite nicely. In more persistent cases, the smoother side of an emery board will do the trick. These are rougher than pumice stones, so use them with discretion.

### *Dubbing Teaser*

In many cases, a fly is rendered more effective by fuzzing up the dubbed body. The old books suggested picking out the dubbing with a needle or bodkin. Tedious. Then, someone recommended burred instruments dentists use for doing root canals. Dangerous; these will cut thread like a hot knife through butter.

I've resorted to a homemade gadget. I simply glue a piece of "male" Velcro to the smoother side of an emery board, using Zap-A-Gap. After the adhesive is dry, I cut the thing into lengthwise strips, to reduce width. This simple tool is one of the handiest on my tying table.

### *Comb*

Fine-toothed combs are great little gizmos for the tyer. As I'll explain later in this book, they're used to remove underfur and short hairs from bunches of deer hair, calf tail, and materials of that sort. They're sold as mustache combs, eyelash combs, and just plain fine-toothed combs. The degree of fineness varies, and it's important to consider what sort of material you'll be combing: Combs that are too fine won't work on coarser hairs and such.

### *Hackle Guards*

These little gadgets are something of a crutch—but if you're limping, a crutch comes in very handy! Their function is to isolate the eye of the hook, holding material out of the way while you apply a whip-finish. They are usually sold in sets of different sizes to accommodate different-sized hooks.

I seldom see hackle guards used by experienced tyers. However, they can be of great benefit to beginners, who have trouble enough learning the whip-finish without having stuff in the way to boot. I still resort to them on occasion when tying Muddler heads, as deer hair has a tendency to get mixed up in the whip-finishing process.

### *Bodkin or Dubbing Needle*

This is a very traditional tool. At one time, it was used for several tasks, including the picking out of dubbed bodies—hence the name. Today, it's more or less relegated to the application of head cement. In fact, I don't even use it for that. I've switched to tiny brushes or—if I need more delicacy—a sharp toothpick, which, in effect, is a throwaway dubbing needle.

If you want a dubbing needle on your tying desk, I recommend purchasing a pin vise at a hardware store. You can mount any type of needle or pin that you desire into this tool and replace it with another when it becomes crusty.

### *Toothpicks*

These are very handy indeed. They serve as throwaway dubbing needles and can be used to apply head lacquer, adhesives, and all sorts of things. Try to find the nice, round ones; they're not as common as they once were.

### *Wing Burner*

While admittedly a rather specialized tool, the wing burner is great for shaping realistic dry-fly wings and such. There are several types available. Your choice will, for the most part, depend on just how sculpted you want your wings to be. Personally, I don't think it makes any difference to the fish that the wings on a dry fly be exactly the shape of those on a natural insect, as long as a believable silhouette is presented. This simplifies matters.

Actually, my wing burner of preference is a homemade device I fashioned out of the thin brass stock you can buy in hobby shops. The wing formers are symmetrical so that by centering the quills, I can burn both wings at once. The design is such that the quills hang out the bottom, which makes it easy to position them.

There's not much more to be said about wing burners. Pick the one that most appeals to you, or make one for yourself. There's not much that can go awry.

### *Head Lacquer or Cement*

While not tools per se, head coatings bear mention here. The most common by far is clear cement. If you're a beginner, I suggest you stay with this for the time being. Later, when you've honed your skills, you may wish to use black or other colors of lacquer, as these can create beautiful effects. We'll look at several of these in this book's streamer-fly section. However, for just plain coating fly heads, clear cement does the job without the potential for making a mess that colored lacquers present.

### *Lamps and Lighting*

It's important to have really good light for fly tying, especially if you're getting along in years and the old orbs aren't what they used to be. Over the years, I've tried a number of lighting devices, some of which have worked out better than others. The three main types, along with their attributes, are:

1. Incandescent light—or, in other words, plain old garden-variety light bulbs. Advantages: inexpensive, many choices, similar to sunlight. Disadvantages: gives off a lot of heat, can take up a lot of space.
2. Fluorescent light. Advantages: cool, restful on eyes, long-lasting. Disadvantages: bulky, somewhat costly, distorts colors.
3. Halogen light. Advantages: similar to sunlight, very bright, very compact, very specular in spotlight configuration. Disadvantages: can be glary, quite pricey.

Despite the high cost, I've switched to halogen overhead lighting in my tying room. I've installed a compact rail-type ceiling fixture that will accommodate up to four sealed-beam spots or floods. I prefer the spots for tying and the floods for lighting the rest of the room, where my supplies are stored. So I have two of each in the fixture. Not only do I have wonderful light, I also no longer have a lamp in the way on my desk.

As I mentioned, fluorescent light is very restful on the eyes and—while not as specular as halogen—can be a very efficient means of lighting a tying area. Installing ceiling fixtures is quite feasible, but if this wouldn't suit the space you have, a good lamp will suffice. Lamps tend to be bulky, however.

Common incandescent light is okay, but it takes a fair amount of it to adequately illuminate a tying area. Either that or the source of light must be quite close to you while you tie, which isn't so pleasant, due to heat and bulk.

If you travel and want to take along your tying stuff, include a lamp; God only knows what you'll find at your destination. There are quite a few lamps on the market, some of which are sold in fly shops. My favorite is a very compact microhalogen model put out by Zelco of Mount

Vernon, New York. It uses a small, 20-watt halogen bulb that provides an amazing amount of light. A pedestal base and clamp are both included. And best of all, the lamp breaks down into small components that can almost be carried in your pocket. An excellent device.

### *Backdrop*

I like to have a neutral background against which to tie. Of course, you can create this in any number of ways, and it can be whatever color you choose. For example, if you're tying Light Cahills, there's nothing like a piece of black velvet to make the pale-colored materials stand out in stark contrast.

If you find yourself becoming a traveling tyer and want a portable backdrop, there's a device on the market that mounts onto tying vises with 3/8-inch-diameter shafts. It's called the Profile Plate and is produced by the Abby Precision Manufacturing Company of Cloverdale, California. It's adjustable for both height and distance and is not overly intrusive—as I had initially feared. The standard plate is a soft white. Additional plates are available in black, pale blue, and pale green; or you can make your own out of matte board, and clip them into place.

This is a very useful tool. It not only provides the desired type of background, but it also reflects a small amount of soft light, which helps bring out definition and detail by backlighting the fly.

### *Vision*

If you think that a vision problem may be compromising your tying, I recommend you take several steps before you go for one of the various magnification setups that are available. Get your eyes checked, and explain to the doctor what you're doing. A cheap pair of diopters from the local drugstore may suffice, as they have in my case. Keep in mind that once you've come to depend on a magnification setup, you're virtually married to it; this can become most burdensome when you're traveling.

As we work through the book, a few other tools will come in for specific mention, including wax, supplementary adhesives, and an interesting device or two. I feel that these are better understood within the context of the tying exercises themselves.

◂ T W O ▸

# Getting Started

As a beginning fly tyer, you'll need to master a few simple techniques involving thread management and use of tools. In the first chapter, you learned how to adjust the vise to accommodate the size of hook you're using and how to mount the hook into the vise jaws. Now you'll begin to learn the basic, essential techniques of fly tying, including how to attach the tying thread to the hook. In the process you'll also learn how to tie four simple, yet deadly, flies that all employ a tyer-friendly material called chenille. More about it in a moment.

The first of these patterns is the famous and ubiquitous Woolly Worm. There are many color and size variations of the Woolly Worm, as there are of most generic patterns. The dressing in this exercise lists the components in the order in which you'll tie them.

Before we start, let me explain some things about materials and components, both in general and specifically for the flies you'll tie in this section.

### About Hooks

Hooks for fly tying come in a very large assortment of sizes, shapes, wire diameters, designs, and today, even colors. Unfortunately, there's no true standardization among the terms used to

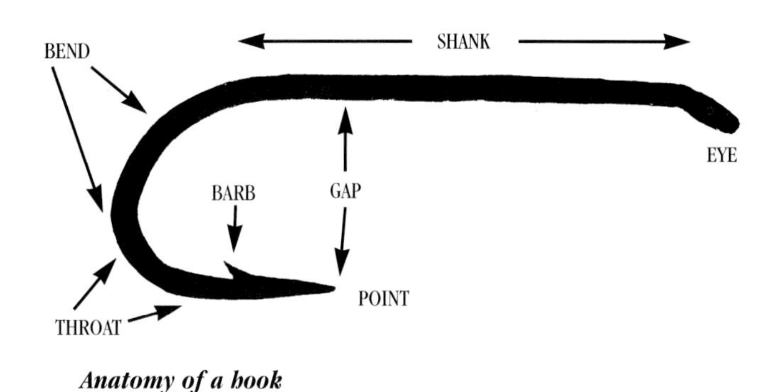


***Anatomy of a hook***

describe such things as shank length, wire diameter, or even size. Thus, it becomes important to know how your finished fly should look and to evaluate the hooks you're about to purchase by those criteria, rather than by what it says on the package. However, I'm going to use the jargon of the trade, because it will make life easier when you're shopping.

With each group of flies we study throughout the book, you'll learn more about the specific hooks required for proper tying. For the flies in this chapter, you'll use hooks with fairly long shanks and medium-heavy wire. These hooks are also suitable for larger nymphs and shorter streamer flies.

Besides size, shank length, and so forth, some hook packages carry a description of the shape of the bend. These terms can also be rather confusing: Model Perfect, Sproat, Modified Sproat, and Limerick are the more common ones. This is what they look like:

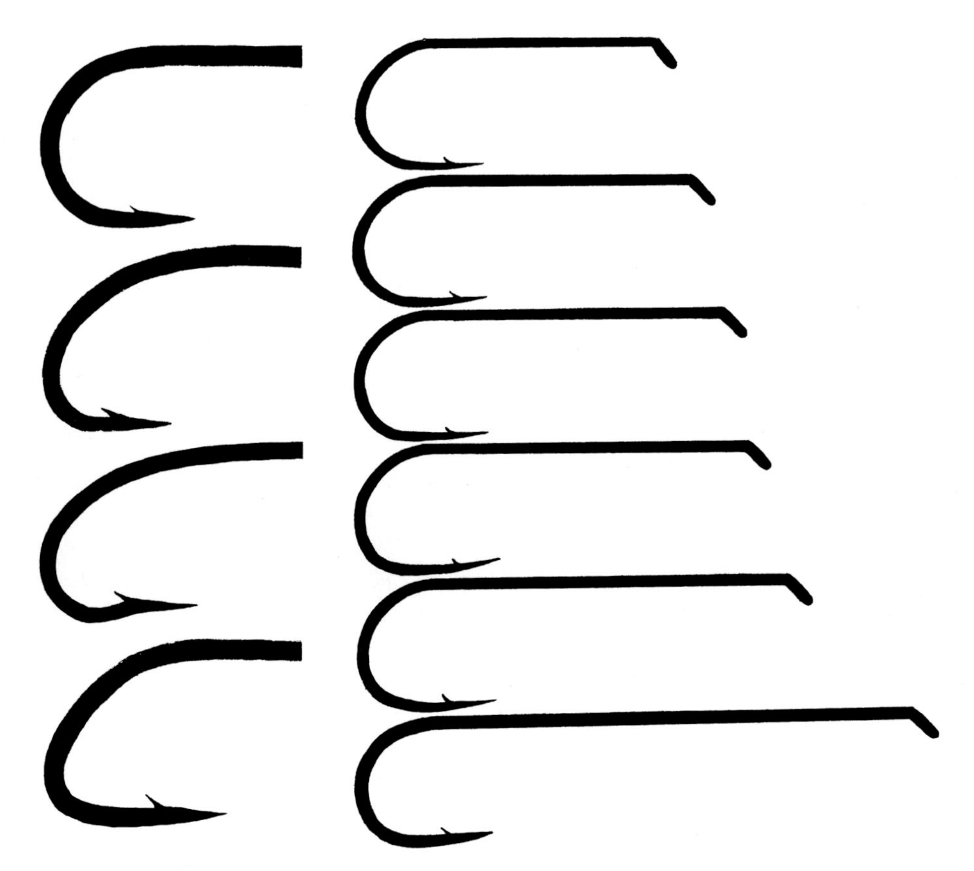

***Hook shapes, or bends. From the top: Model Perfect, Sproat, Modified Sproat, Limerick.***

***Relative shank lengths. From the top: standard, 1X long, 2X long, 3X long, 4X long, 6X long.***

Model Perfect simply means that the bend describes a semicircle. This design is best for dry flies, as it sets the tail into the desired position. Model Perfect hooks can also be used for other types of flies; however, I prefer different bend shapes, as they position the hook point where I want it and balance out the symmetry of the fly in a more graceful manner.

A few words about hook design and determining the best one for your needs. The eye shouldn't slope downward at too extreme an angle; this causes problems both in finishing the head of the fly and, later on, in tying it to the leader. About 30 degrees is good. There are also straight-eye hooks. These are fine for certain types of flies, but keep in mind that several knots commonly used to tie on flies, notably the Turle and its variations, are not accommodated by a straight eye.

There are also hooks with turned-up eyes. To a large extent, they fall within the province of the salmon-fly tyer, and we won't be addressing them here.

Another important consideration with regard to hook eyes is how well they are formed.

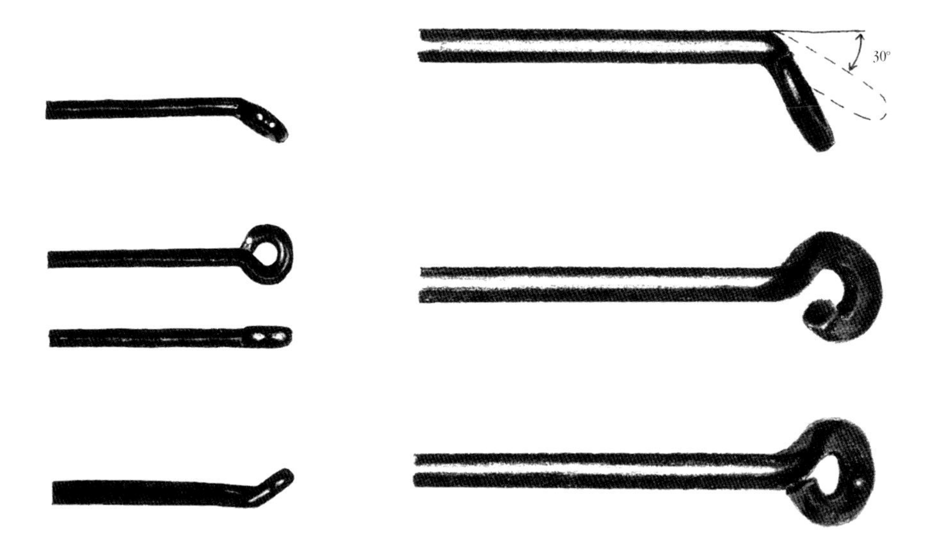


***Hook eye types. From the top: turned-down; straight, or ringed (top and side views); turned-up.***

***Potential problems with hook eyes. Top: extreme turn-down, in excess of the 30 to 35 degrees that is recommended. Center: an open eye with sharp-edged wire exposed. Bottom: a well-formed eye.***

The eye should be closed—that is, there should be no gap at the spot where the wire loops back to the main shank. Also, there should be no roughness there. Some cheaper hooks are delinquent in these areas. Do yourself a good turn and avoid them.

Hook points and barbs likewise come in for evaluation. Today, we release our fish, and a smaller barb better facilitates this. Also, such a design better accommodates pinching down the barb without breakage.

The shape of the hook point is closely related to the thickness of the wire. Very fine wire hooks, such as dry-fly models, generally have fairly short points, because these penetrate more easily. Heavier-wire hooks require a more tapered point for optimal penetration and also to allow resharpening without stubbing down the point too much.

Dry-fly hooks also have—or should have—very conservative barbs. It's quite difficult for fish to dislodge even debarbed light-wire hooks; they have little leverage or weight to work against. When pinching down barbs, be as gentle as the task will allow, and do it before you tie the fly, in case breakage should occur.

I happen to have a preference for Daiichi hooks. This is not to disparage other makes; there are many good ones out there. In the tying exercises, I'll list the Daiichi model number and description. The appendix at the back of the book is a hook substitution chart that will enable you to select similar hooks from other makers.

## *About Threads*

Today's threads are so much better that those I started with back in 1959 that words are hardly adequate to describe the differences.

The major benefits are:

- Modern threads are much stronger in relation to their thickness than were the older ones; thus, there's less bulk per wrap. More wraps with finer thread produce better results than fewer wraps with thicker thread.
- Being synthetic, shelf life is not an issue.
- They have high resistance to fraying.
- Lots of useful colors are available.

Two very popular threads in use today are 8/0 Uni-Thread and Danville Flymaster, which is rated at 6/0. These numeric designations may have had a specific meaning at one time, but now they are simply indicators of relative thickness.

The two threads are somewhat different. While slightly finer, the Uni product, which is made of Dacron, is the stronger of the two and the more pleasant to work with. The Danville product, being nylon, lies flatter; some tyers who do artistic tying like this attribute. My preference is for the Uni, but I'll leave the choice to you. I can assure you that new threads are coming onto the market with increasing frequency, as I've indicated in my *Modern Fly-Tying Materials.*

Some threads, Uni in particular, offer a choice of waxed or unwaxed. With Uni, either will work just fine. The material of which the thread is made does not require wax to keep the fibers together. Wax is used only to help control slippery materials. Danville is more slippery and definitely requires wax. If you do opt for waxed thread, be sure that your head cement is fairly well thinned so that it will penetrate and bond.

### *About Chenille*

*Chenille* is the French word for "caterpillar," and when you see the stuff, you'll realize the appropriateness of the name. It consists of fuzzy material with a thread core. It comes in a wide assortment of colors and four or five thicknesses. It's important to take the latter into consideration when you select chenille, as proper diameter relative to fly size is critical.

There's also a quality consideration. Look for chenille that's relatively soft and fine of texture. It wraps better, looks nicer, and accommodates hackle more readily.

I've seen another type of chenille under the names Ultra-chenille and Vernille. Here, the fuzzy material is bonded to a monofilament core. It's somewhat less flexible than regular chenille and accommodates certain special applications, such as making the infamous San Juan Worm.

### *About Saddle Hackle*

The term *hackle* is widely encountered in fly tying. It refers to the material—almost always a feather of some type—that's found at the throat, and sometimes over the body, of flies. In this case, I'm talking about the saddle hackle used for larger wet flies, such as the ones we're about to tie. We'll look at dry-fly saddles farther on.

Saddle hackle, as differentiated from neck or cape hackle, comes from the back, or saddle, of a chicken. The kind we'll be using here is longer, softer, and webbier than cape hackle and features finer quills. For Woolly Worms and such, these are just the attributes you want. Keep that in mind when shopping.

### *Hackle Nomenclature*

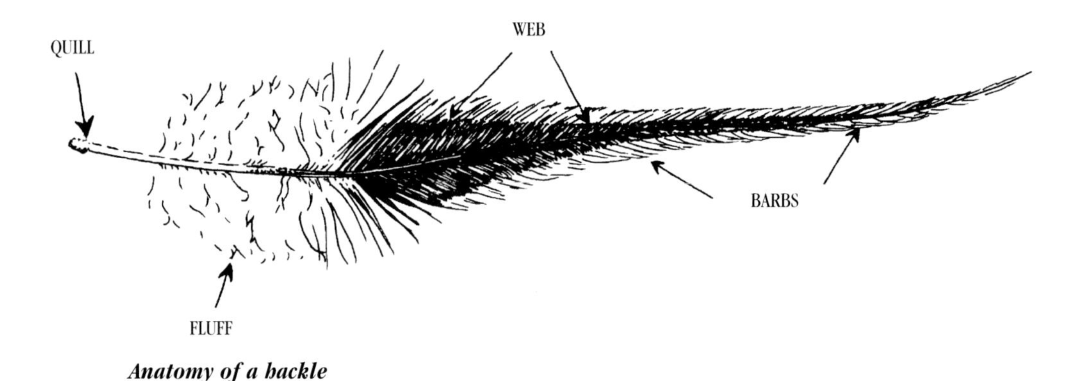


***Anatomy of a hackle***

## Tying Terms

> **Note:** The following descriptions of tying procedures assume a right-handed tyer. If you're tying lefty, simply transpose the terms **right hand** and **left hand.**

***Thread Torque*** This refers to the tendency of thread to push materials ahead of itself during wrapping. There are circumstances in which you can use thread torque to your advantage by allowing it to distribute material around the hook in a controlled manner. The most common application of this is when you're spinning hair, as in Muddler Minnow collar/head assemblies. Most of the time, however, you'll use thread management techniques to counteract thread torque.

***Pinch-Wrap*** This technique is used to affix materials precisely where you place them, without thread torque pushing them around the hook. Hold the material in position with your left thumb and forefinger. Sneak up the thread between your thumb and forefinger on the near side of the hook, then down between them again on the far side of the hook. Then, tighten the thread while your left hand holds everything in position. Usually, several pinch-wraps are applied in succession.

***Soft-Wrap*** This technique is easy to execute and can be used in place of the pinch-wrap when the material you're tying in is resistant enough that it won't readily react to thread torque. Simply relax your bobbin hand's tension on the thread and "sneak" it over the material. Gradually increase tension with succeeding wraps to secure.

***Lift-Over Wrap*** This move involves switching hands. It's most commonly employed when you're tying off a piece of material that would compromise working room if you used the normal thread-wrapping method. The right hand holds the material under tension. The left hand manipulates the bobbin by passing it over the top, thus making a thread wrap over the material. The left hand then releases the bobbin, catches it again beneath the hook, and repeats the move as many times as is required to secure the material.

***Palmered Hackle*** This describes hackle that is wrapped over all or part of the fly's body, rather than only near its front.

***The Dressing*** Olive Grizzly Woolly Worm

*Hook:* Daiichi #1720 or #2220; 3X- or 4X-long, respectively. The #2220 is actually a short streamer-fly hook.

*Hook Size:* For this exercise, size 6 or 8.

*Thread:* Fine diameter, such as 8/0 Uni-Thread or comparable; black.

*Body:* Medium-thick chenille; olive.

*Hackle:* Grizzly (barred rock) saddle hackle; long and fairly soft.

## Tying Steps

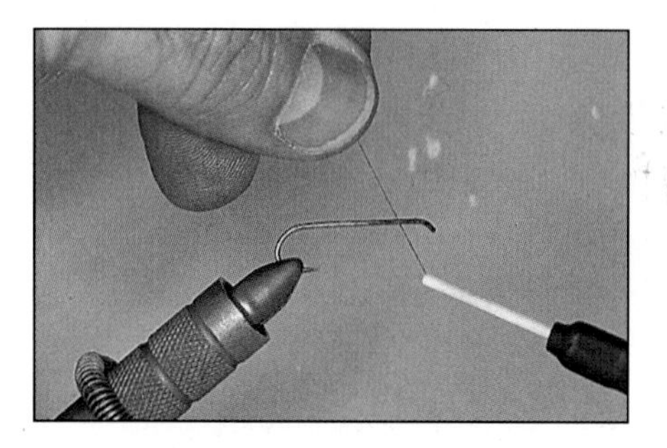

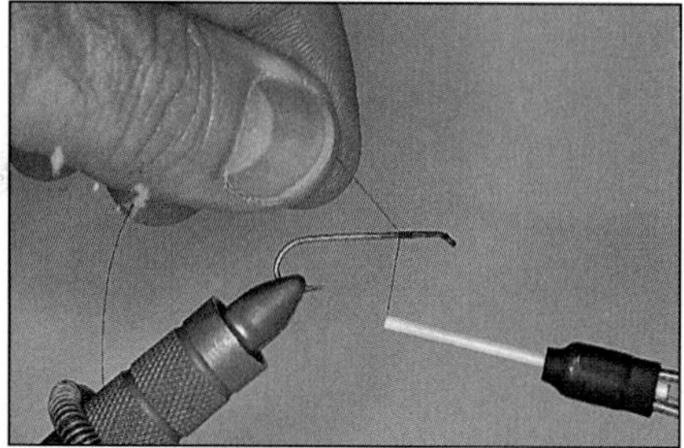

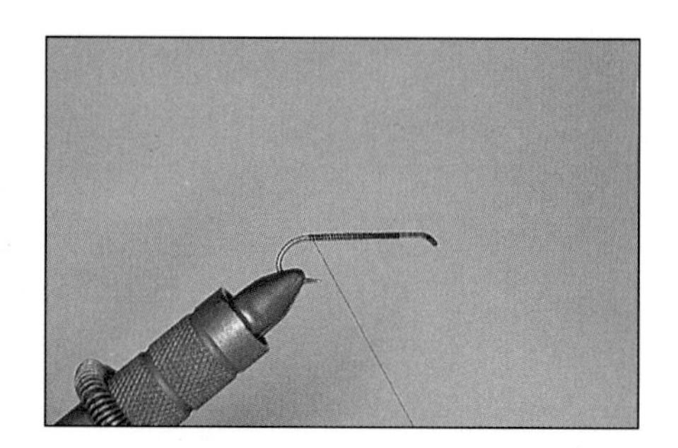

***Attaching thread to hook; steps 1-5.***

1. Secure the hook in the vise jaws, making sure that the shank is on a flat plane. pull a few inches of thread out of the bobbin. Take this thread in your left hand and hold it beneath the hook shank, a little way back from the eye. Your left hand is on the far side of the hook. Your right hand is holding the bobbin and is on the near side of the hook.
2. While holding the thread under tension, begin to wrap the thread back over itself toward the rear of the hook. *Wrap over and away from yourself.* In fact, almost all fly-tying operations move in that direction. There are a few exceptions, but don't be concerned about that right now. Learn to work with only a moderate amount of thread exposed from the bobbin tube, so that you aren't making large, awkward wraps. In this case, it isn't necessary for the wraps to be contiguous, although there will be times when this is desirable.
3. After you've taken perhaps six turns, let the bobbin hang; its weight will maintain the thread tension. Hold the tag end of the thread tightly with your left hand and, using your scissors as a lance, cut it off flush with the hook.
4. Continue to spiral-wrap the thread rearward. As you near the bend, be careful to swing the thread around the point of the hook with each turn, so as to avoid abrading or cutting it.
5. When you've reached the bend—that is, the spot at which the hook begins to slope—stop.

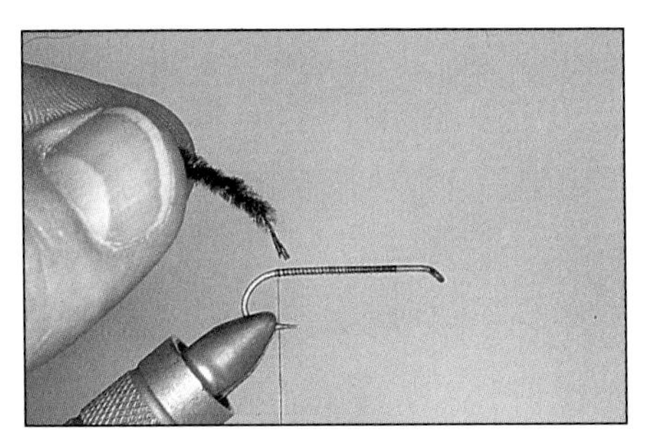

***Chenille core exposed; step 6.***

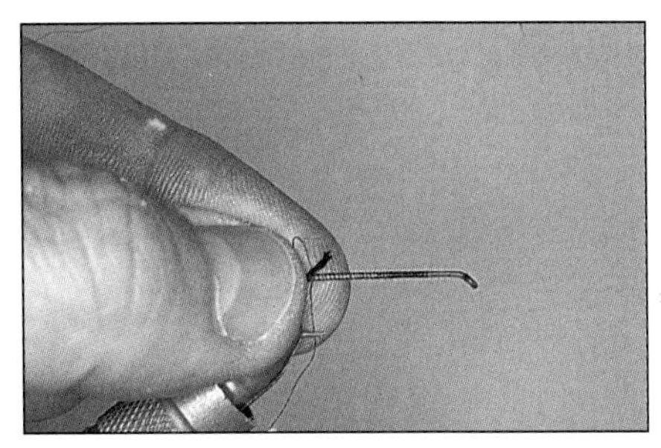

***Tying on the chenille; steps 6-8.***

**6.** Cut a piece of chenille about 7 inches in length. Manicure it as follows: Scrape off a little of the fuzz from one end, exposing about 1/4 inch of the thread core.

**7.** Tie in the chenille by this exposed core at the rear of the hook shank, just at the point where the bend begins. This technique eliminates the bulk that would result if you tied in the chenille en masse. You can use either the pinch-wrap or the soft-wrap technique for this—whichever works for you.

**8.** Wrap the thread forward a few turns—a distance approximately equal to one thickness of chenille—and let it hang there. This will allow you to take one wrap of chenille behind the feather, thus protecting the quill at that spot.

***A saddle hackle; step 9.***

***Saddle hackle, with barbs flared; step 9.***

**9.** Choose a saddle hackle that has barbs about 1 1/2 times the gape of the hook in length. Gently, so as not to tear off the barbs, stroke toward the butt end, so that the barbs stand out at an angle from the quill. This maneuver is easier if you start at the butt end.

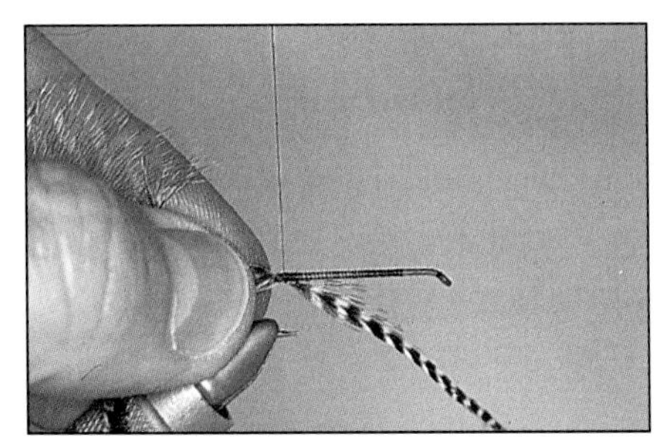

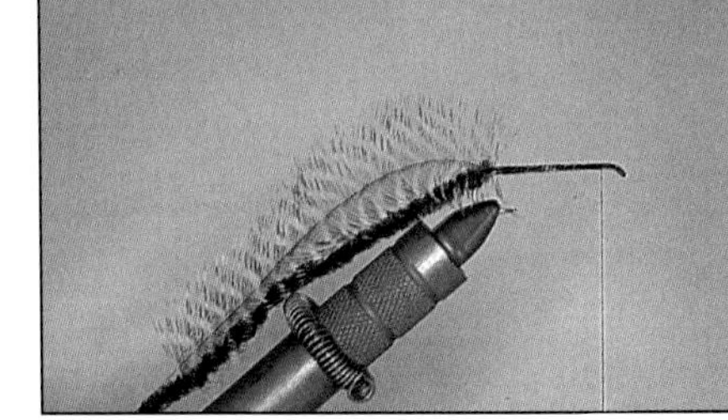

***Tying in and securing the saddle hackle; steps 10-11.***

**10.** Fold back the barbs near the tip end of the feather, exposing the quill in that area. With a few soft wraps, tie in the feather by that end, where the thread was left hanging. Be sure to tie it in "pretty-side-forward": in other words, with the exterior side of the feather, as it was positioned on the bird, forward.

**11.** Secure the feather by wrapping forward over the tag end with firm tension. Trim, if necessary. Continue wrapping forward until you reach the front portion of the hook, but be sure to leave an adequate amount of space to tie off the chenille and the hackle and to create a whip-finish.

*Wrapping the chenille; step 12.*

12. Pick up the chenille and make the first turn behind the tie-in point of the hackle feather. Then, with the next turn, come in front of the feather, and continue wrapping forward under fairly firm tension until you reach the point where the thread is hanging.
13. Tie off the chenille with a series of firm thread wraps. This is done by holding the chenille under tension with your right hand and executing several or more "lift-over" wraps: passing the bobbin over the top, letting it go, then catching it again from underneath.
14. When you're sure the chenille is secure, neatly trim off the excess. Then take a few turns of thread over the trimmed spot to neaten things up. Maintain some clearance from the eye.

*Tying off and trimming the chenille; steps 13-14.*

15. Pick up the hackle feather and spiral-wrap it forward. Try to follow the grooves between the turns of chenille, and allow the quill to sink into the soft material. Make sure the pretty side is in front. Keep the quill straight, but be gentle; this is no time for the quill to break. If the barbs tend to bind, stroke them rearward a little with your left hand. Depending on the length of the feather, you may or may not need your hackle pliers—or at least not until you have wrapped most of the way forward.

**Note:** If your feather proves to be too short for comfortable and secure gripping with fingers alone, by all means use hackle pliers. The recommended way to grip the feather is straight-on by the tip of the quill, so that everything is kept in a straight line, and there's no back-and-forth flexing of the quill.

16. This is optional, but advisable: As you reach the end of the body, take two or three turns of hackle directly in front, abutting the chenille. While making these turns, try

to stroke the barbs back a little. This gives the fly a nicely finished appearance and, I believe, makes it fishier.

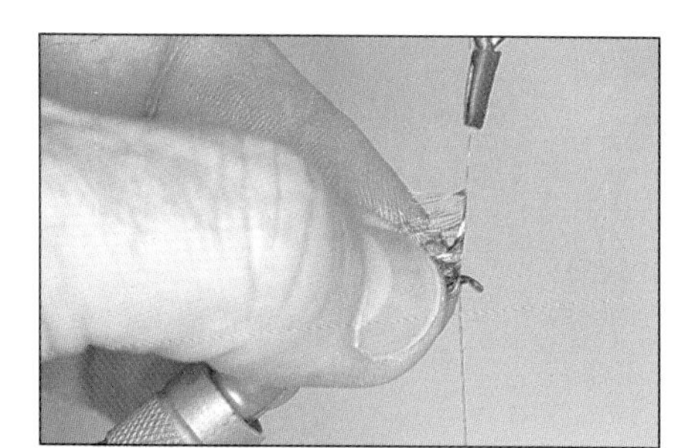

***Wrapping the hackle; steps 15-16.***

***The hackle is tied off by binding down the quill. When cutting off the excess quill, hold it to the side with hackle pliers, as shown, to avoid cutting the thread; step 17.***

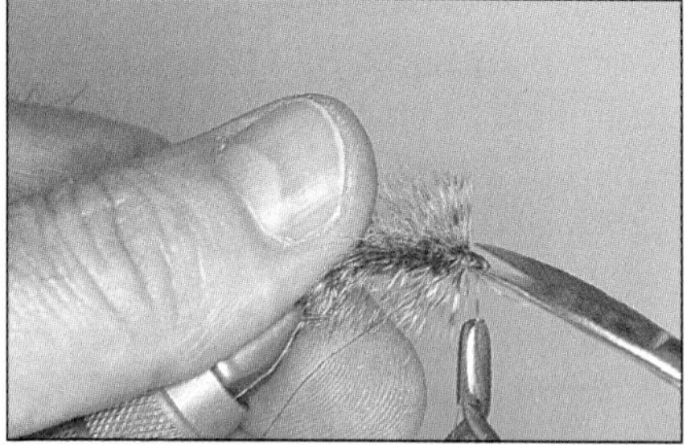

***During any trimming of unruly barbs, hold the thread out of the way.***

***The finished fly.***

17. Tie off the hackle feather as you did the chenille. When you've firmly secured it, trim off the excess, take a few "housekeeping" thread wraps, and create a whip-finish. I advise adding a coat or two of clear head cement for the sake of both durability and appearance.

Is this the first fly you've ever tied? Congratulations! I suggest you put it in a box with a note to that effect. I sorely wish I had saved my first fly ever—in fact, a selection of my early efforts.

Now that you can tie the Woolly Worm, it's a simple matter to add a marabou tail, converting the fly into the redoubtable Woolly Bugger. I'll also show you how to weight the body with lead wire.

### *About Marabou*

Marabou is one of the most popular and ubiquitous fly-tying materials. Originally, this material came from a huge, ungainly bird indigenous to parts of Africa called the marabou stork. This bird is now protected, and today's marabou is actually turkey body feathers.

Marabou is a very soft, absorbent material. Its main use is on streamer flies, which are actually baitfish imitations. When wet, it undulates and pulsates in the current, which results in a suggestion of aliveness. Predator fish find this most appealing.

Marabou comes in little bundles, with the feathers strung together by the butt ends. Be aware that these feathers may differ widely in size and shape; you should look them over before purchasing. You'll encounter terms like marabou *shorts* or *bloods.* Actually, some of these shorts are long enough to tie a sizable streamer. Conversely, some may be so short as to be suitable only for modest-sized Woolly Bugger tails. So inspect the bundle, and be sure it contains the size of feathers you want. If you're buying from a catalog, tell the supplier what you're going to be tying. These days, most of the better retailers have knowledgeable tyers on their staffs.

There are a couple of ways to work with marabou. These will be covered in the tying exercise.

### *About Lead Wire*

For many years, American anglers have used weighted flies and have done so guiltlessly. This was a virtual no-no on English chalk streams and, in many of these venues, still is. Even William Lunn, the famous angler/riverkeeper, was castigated by some for using fine copper wire when tying his lethal nymphs.

By far the most commonly used material for weighting flies is lead wire. It's cheap and easy to use—as you'll see in a moment. It comes on spools and in a number of diameters, ranging from 0.010 inch, the thinnest practical wire, to .035 and .040, which are pretty stout and are appropriate only for very large flies, such as those used in Alaska.

From a tying standpoint, the thicker the wire, the more difficult it is to work with, so I tend to be conservative. When making wrapped-wire underbodies, I rarely use a wire diameter larger than .020 inch, even on my Alaska pool-dredgers. In extreme angling circumstances, you can always add a split shot or two.

Those of you who hunt with shotguns are aware that lead shot has been banned, at least from waterfowl hunting. I think this had to be done. Lead is unquestionably toxic; over time, the bottoms of marshes had become receptacles for countless billions of pellets. The birds were ingesting them, with lethal results. So the ban is justified.

In recent years, there have been some noises to the effect that lead should be banned in fly fishing and tying, as well. In fact, nontoxic split shot is now available, and lead in any form has been banned from the waterways of the national parks. I'm not sure that this makes much sense, given the small amount of lead fly fishing introduces into the environment. However, we've been stridently proselytizing the politicians over environmental issues for years, and we may have to bite our own bullets (no pun intended ) on this one. We shall see. I should hope that, technology being seemingly infinite in scope, a suitable replacement for lead wire can be developed.

And now, the Woolly Bugger.

***The Dressing*** Woolly Bugger

*Hook, Hook Size, Thread:* Same as Woolly Worm.

*Tail:* Black marabou.

*Underbody:* Lead wire, .015" or .020" diameter.

*Body, Hackle:* Same as for Woolly Worm.

**Tying Steps**

1. Start the thread the same way you did for the Worm. Let it hang at the rear, and prepare a clump of marabou by one of the two following methods:
   (A) If the shape of the feather allows, simply pull off the fluffy material near the butt

end. Stroke the fibers rearward, bunching them, and tie in the feather whole. You may need to snip out the quill if it extends too far toward the tip.

(B) With larger plumes that have heavier quills, stroke the fibers outward to a 90-degree angle from the quill and cut them off in bunches. You may need several clumps, depending on the size of your fly.

2. The standard length of the tail is equal to the length of the hook shank. Later, you can vary this in either direction as you wish, but for now let's adhere to the norm. Gauge the proportions by laying the marabou alongside the hook. Then tie in the bunch directly on top of the hook at the bend, using a series of pinch-wraps.

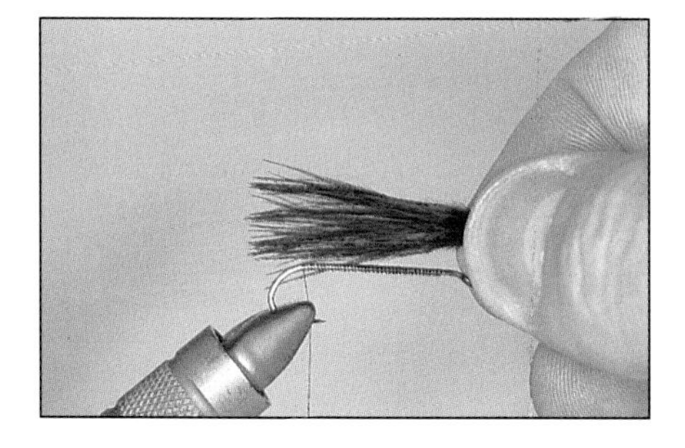

***Quantity and length of marabou tail; steps 1-2.***

***Securing and trimming the marabou; steps 2-4.***

3. Secure the marabou with a series of firm, neat wraps, working the thread back and forth over a short area, as shown in the photograph. Then trim the excess off flush, being careful not to cut the thread in the process. Let the thread remain at the rear.
4. Cut off a piece of lead wire about 8 inches in length and simply wrap it around the hook, starting where the marabou ends. The wire should abut the "step" that was created by the trimming. Wrap the wire forward to a position just short of where you'll want the front of the body to extend, and cut off the excess at both ends. Allow enough space for one or two turns of chenille in front of the lead; this creates a neat front taper and sets up easy completion of the fly.
5. Optional, but advisable: coat the wire wraps with clear head lacquer or—even better—a superglue, such as Zap-A-Gap. The adhesive will run down into the little crevices and help bond the wire to the hook.

***Applying the lead wire; steps 4-5.***

6. From here on, the Bugger is tied the same as the Woolly Worm, except that after tying in the chenille and hackle you'll be spiral-wrapping the thread forward over the lead wire.

***Completing the Woolly Bugger; step 6.***

***Another Woolly Bugger, very useful as a damselfly-nymph imitation.***

**Note:** Observe in the photos that the presence of the wire enlarges the thickness of the fly's body. No problem, except that you must be sure to choose a saddle hackle long enough to compensate for this.

Also, I should note that Woolly Buggers can be tied in different styles. The brown-and-olive one in the accompanying picture is tied with slightly webbier hackle and a short tail. It has served me very well as a damselfly nymph imitation. Brightly colored, short-tailed Woolly Buggers in small sizes make excellent panfish flies.

Now let's implement our knowledge of how to work with chenille by incorporating it into a fly that employs yet another easy-to-use, low-cost material. I'm referring to rubber legs.

## *About Rubber Legs*

This material, when separated into strands, looks like thin rubber bands. It imbues flies with a most alluring sort of aliveness, and quite a number of dressings call for it, particularly Rocky Mountain patterns.

I well recall my first encounter with rubber legs. It happened in Dan Bailey's shop in Livingston, Montana, back when the original proprietor was still alive. It was my first trip west, and I was intrigued by the regional flies on display. I asked about a particular pattern and was told that it was called the Girdle Bug, because the rubber leg material was procured from lady's girdles of that era. I bought a bunch of the flies, and they turned out to be terrific fish-catchers, much to the dismay of my companion, who was and is very much a dry-fly advocate.

Rubber leg material comes in die-cut sheets. The individual strands are easily separated with the fingers or with the aid of a tweezer. For the thicker legs on larger flies, you can use a double strand. A selection of colors is available.

***The Dressing*** Girdle Bug

*Hook:* Daiichi #1710 or #1720; 2X- or 3X-long, respectively.

*Hook Size:* For this exercise, size 6 or 8.

*Thread:* 8/0 Uni-Thread or comparable; black or to match body.

*Body:* Fine or medium-fine chenille; color to suit yourself.

*Underbody (optional):* Fine lead wire (not used for this exercise).

*Legs:* White rubber leg material.

## Tying Steps

1. Begin the Girdle Bug as you did the Woolly Worm; stripping and tying in the chenille at the rear. Advance the thread about 1/4 of the way up the hook shank.

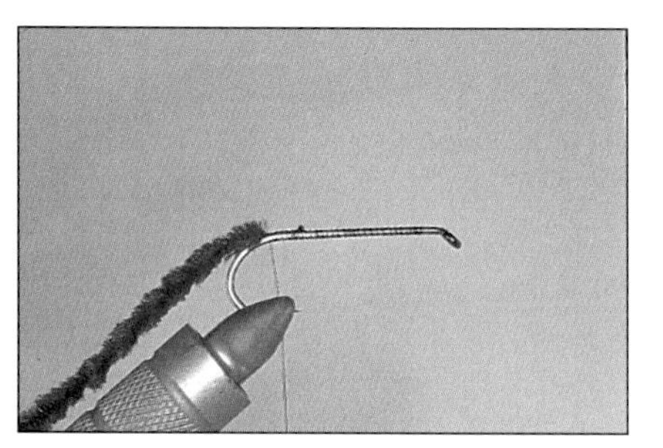

***The Girdle Bug starts out like the Woolly Worm; step 1.***

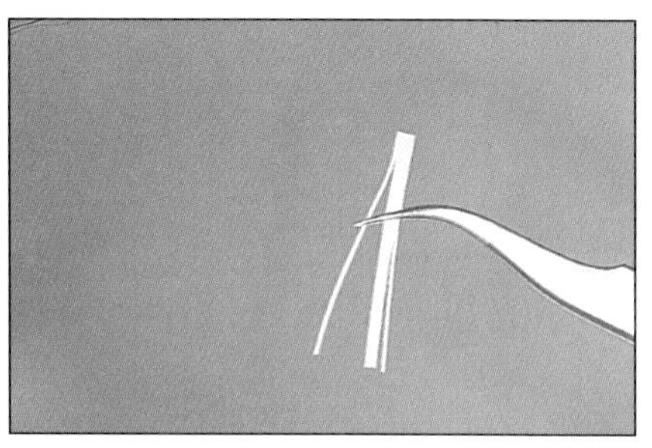

***Separating rubber leg material into strips; step 2.***

2. Prepare three pieces of rubber leg material by separating them into single strands and cutting each to about an inch in length.
3. Tie in the first piece as follows: Center it on the hook shank, and secure it with two pinch-wraps. Then hold the near end with your right hand and make two lift-over wraps, handling the bobbin with your left hand (as described in the Woolly Worm exercise). These wraps cross over the other two in X-wrap fashion, which will cause the rubber leg material to be repositioned at a 90-degree angle to the hook shank. Secure with a few more X-wraps, crisscrossing with each wrap.

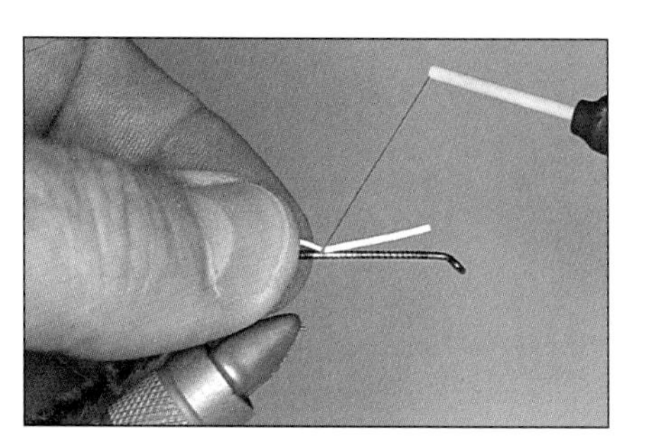

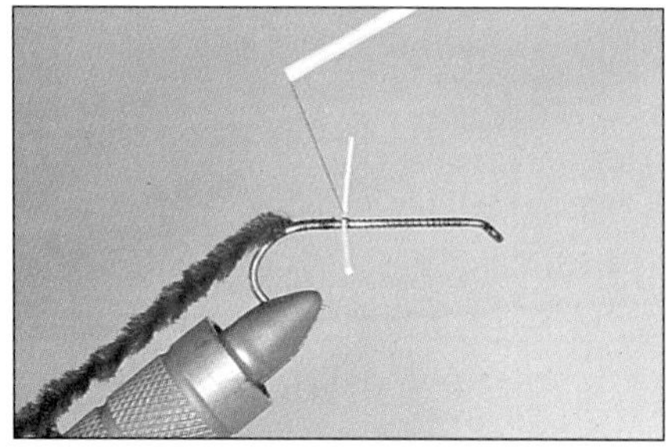

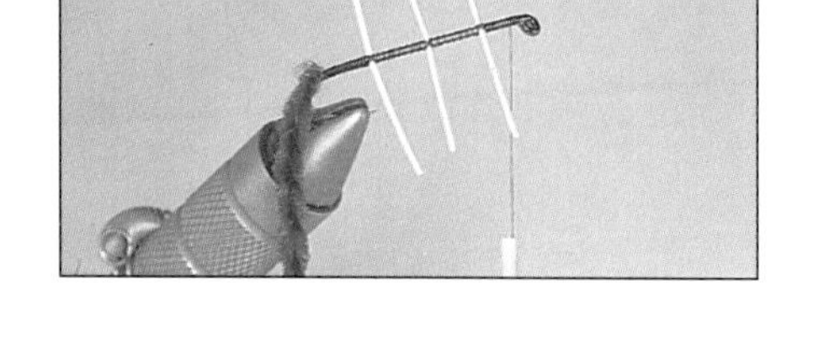

***Tying on the rubber legs; steps 3-4.***

4. Make two more sets of legs in exactly the same manner, tying one on at midshank and the other at about the 3/4 point, spaced evenly.
5. I won't repeat the instructions here, but if you want to weight the body, do so just as you did the Woolly Bugger, except working the wire between the sets of legs. In this case, the thread is up front, having been used to tie in the legs. If you find that this interferes with the weighting, simply tie off with a three-turn whip-finish, wrap the wire, and tie back on again.

*Wrapping the body; step 6.*

*The completed fly.*

6. Wrap the chenille, working around and between the legs. Tie off and trim the chenille up front, whip-finish, and lacquer.
7. If necessary, adjust the length of the legs, trimming them so that they're equal.

Let's close with one more pattern that should be a cinch for you, because it incorporates materials and techniques with which you're already familiar. The only differences are the design and shape of the fly; the hackle; and the way in which you handle the primary material, chenille. This is an elongated fly that effectively imitates the nymphal form of larger stone flies and is a terrific general pattern as well. I've seen it tied in various colors, the most common being black and brown. We'll tie a brown one here.

## Tying Terms

***Thorax*** This component bears the same name as, and corresponds in location to, the thorax of a real nymph.

***Wing Case*** This again correlates in name and position with a body part of a real nymph: the dark casing on the back that protects the budding wings until the insect is ready to hatch.

***The Dressing*** Montana Nymph

*Hook:* Daiichi #1720 or #2220; 3X- or 4X-long, respectively.

*Hook Size:* For this exercise, size 6 or 8.

*Thread:* 8/0 Uni-Thread; black or dark brown.

*Tail:* Webby brown hackle barbs.

*Underbody (optional):* .015" lead wire.

*Body/Wing Case:* Fine to medium chenille (see note); brown.

*Thorax:* Fine to medium chenille; deep yellow or light orange.

*Hackle:* Brown, fairly soft; preferably webby rooster saddle.

### Tying Steps

1. Tie on a little way back from the eye and wrap to the bend.
2. Please look ahead to chapter 3 and familiarize yourself with the barb-gathering and tailing procedure covered in the Iron Blue Dun and Dark Cahill tying sequences. Using this method, tie on a fairly short tail, trim the butts, and cover with thread, as shown.
3. At the bend, tie in a 6- or 7-inch piece of brown chenille. Remember to prepare it by stripping off a little of the fuzz, so that you can tie it in by just the thread core.

4. If you've opted for weight, wrap the lead underbody at this time, using the method described for the Woolly Bugger. Stay well back from the eye.

5. Advance the thread $^{2}/_{3}$ of the way up the shank and let it hang. Pick up the chenille, wrap it forward to where the thread is waiting, and tie it down, preferably on top of the hook, for reasons that will become apparent with the next step. *Don't* cut off the excess!

6. Make a good-sized loop with the chenille, as shown in the photo. Tie it down right where you tied down the body, on top of the hook. There are now three strands of chenille: the loop and the tag. These will be brought over the top of the thorax to form the wing case. For now, fold them up out of the way as best you can. If the tag end is long enough, you can catch the loop beneath it and secure it in your materials clip; this will clear an area for you to work on the thorax and hackle.

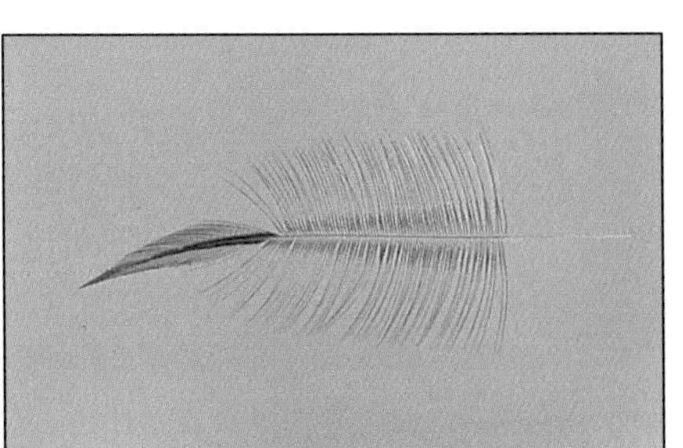

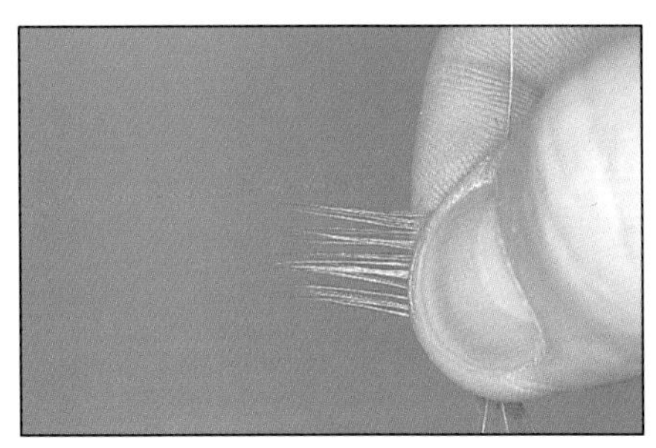

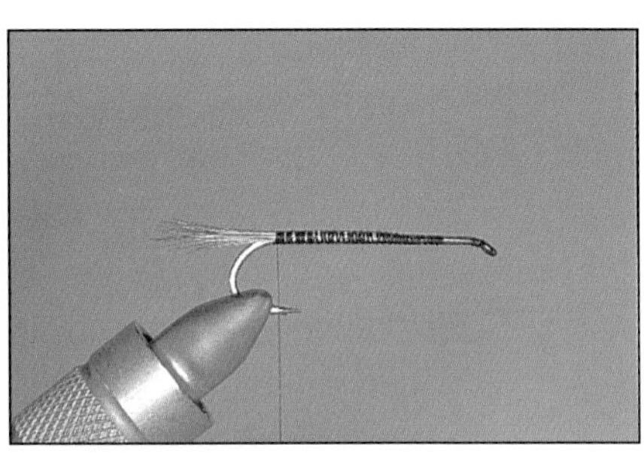

***The tailing procedure. Barbs stroked to a 90-degree angle to the quill, barbs gathered into a bunch, tail tied in place, butts covered with thread; step 2.***

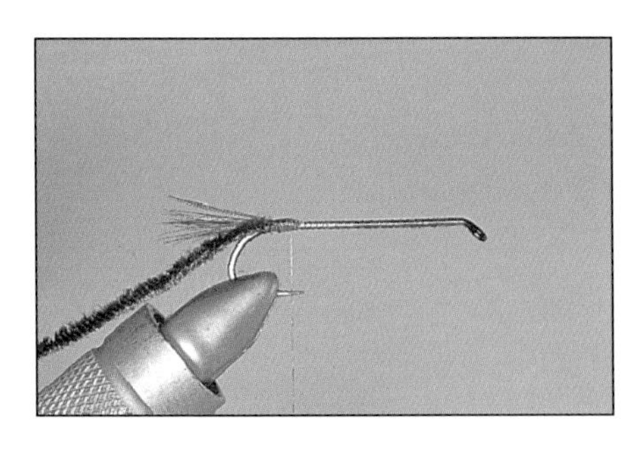

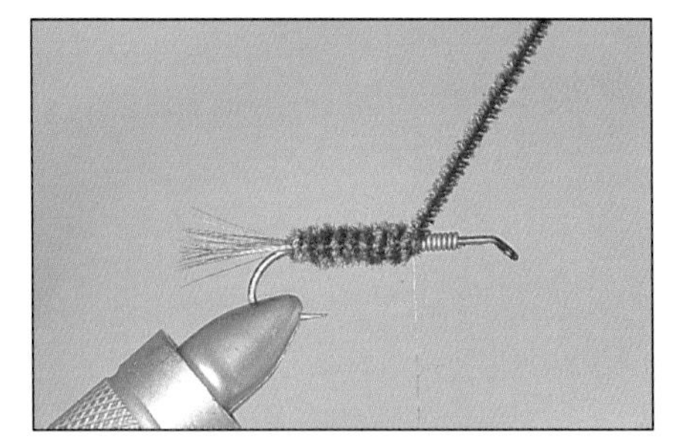

***Applying the lead wire and wrapping the body; steps 3-5.***

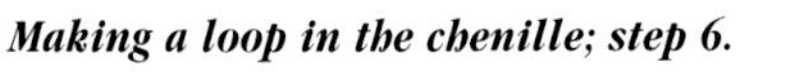

***Making a loop in the chenille; step 6.***

***Attaching throat chenille; step 7.***

***Hackle feather tied in; step 8.***

7. Prepare a 2- or 3-inch (50–75mm) piece of the yellow or orange chenille as you did the brown, and tie it in at the front of the body.

8. Prepare a hackle using the same method as for the Woolly Worm, but this time tie in the hackle by the butt end, pretty-side-forward, leaving a tiny length of bare quill exposed. Trim off the quill butt.

***Throat chenille wrapped in place and trimmed; step 9.***

***Hackle wrapped and secure; step 10.***

***Bring the loop and tag end of chenille over the top, and tie them down; step 11.***

9. Pick up the yellow or orange chenille and take several turns, working forward. *Impor- tant: Don't come anywhere near the eye!* When the wing case material is folded over, there will be a lot to tie down, and you'll need plenty of space. Tie off and trim.
10. Wrap the hackle around and through the thorax chenille, Woolly Worm style. Tie off and trim.
11. Gather up the three strands of brown chenille and bring them over the top, thus forming a prominent wing case. Tie down securely, trim, whip-finish, and lacquer.

***The completed Montana.***

**Note:** The only difficulty with the Montana Nymph is dealing with the bulk when tying off. For this reason, you shouldn't use too heavy a chenille for the size of hook you're working on. If you find that you're still having trouble, trim off the tag end, and use only the two sides of the loop to form the wing case; they're sufficient. Another solution is to use a straight-eye rather than a turned-down-eye hook; this will make the tie-off easier.

Believe it or not, you're now prepared to tie quite a number of highly effective fly patterns. Practice your technique by tying a few more Worms, Buggers, Bugs, and Montanas, and you'll be ready for the next lesson.

◂THREE▸

# *Wet Flies*

In this chapter you'll learn about the various methods for making bodies, wings, hackles, and tails, along with an embellishment known as *ribbing* for subsurface flies. You'll work with several types of furlike materials and also a stranded material that's wrapped in a manner somewhat similar to chenille, but isn't quite so forgiving.

## Tying Terms

***Dubbing*** This word is both a noun and a verb; in other words, it's a material, and it's also the handling of that material. Basically, dubbing is fur or furlike material. It varies widely in texture, and thus in application, as you'll see. It may be natural fur, synthetic fur, or a mixture of the two.

Dubbings are commonly referred to as being *smooth* or *rough.* These terms describe two rather distinctive types, and there are many textures between the two extremes. It's important to understand what sort of dubbing suits what type of fly, and I'll devote considerable attention to that throughout this book.

A particular type that you'll be using shortly is called *hare's-ear* dubbing. It dates back to early British fly tying. The original material was described in old books as "poll scraped from the ears and mask of a hare." This is still used, but today other furs, such as squirrel, are often blended in to achieve a similar effect.

You'll also hear the term *guard hair;* this is the coarser, harder exterior hair that gives an animal's coat its shine and protects it from contact with brush, or whatever. When guard hair is included in a dubbing mix, the material is often referred to as being *spiky.*

***Floss*** This is not the stuff you're supposed to use to scrape the tartar off your teeth. It's stranded, spooled material for making fly bodies. It's wrapped like chenille, but it doesn't hide what's underneath it the way chenille and heavy yarns and dubbings do, so it requires a different technique. Floss comes in a wide array of colors and a selection of thicknesses. Several materials are used for making floss, including real silk, which is still preferred by traditionalists. My favorite floss is made of rayon, but who knows what might be coming?

***Tinsel*** This is not the kind that adorns small pine trees in December. It's a shiny, metallic sort of material; however, the most common type of tinsel in use today isn't truly metal, but rather a space-age product called Mylar. The three most popular colors are silver, gold, and copper, but there are others. Tinsel comes in two basic forms: *flat* and *oval.* Four or five widths of the flat type are available, which facilitates tying flies of all sizes. The oval type has a fine fiber core and also comes in a variety of thicknesses. It's mainly used as ribbing.

***Ribbing*** When used as a verb, this term refers to wrapping supplemental material over the body of a fly, usually in spiral fashion, in order to add a highlight or to change the character of the body. The most frequently used material is tinsel. For ribbing softer, fuzzier bodies, such as those made of soft yarn, dubbing, or peacock herl, I prefer oval to flat tinsel. It ties better, looks better, and is stronger. Flat tinsel works beautifully over floss and similar materials, as you'll see on the Grizzly King in chapter 7.

***Flank Feathers*** These are feathers from the sides of game birds, mainly waterfowl. They grow beneath the birds' wings and are widely used for making wings on both wet and dry flies. The most desirable is *wood duck* flank; it's also the most expensive and is generally in short supply. *Mallard* and *teal* flank are commonly dyed to resemble wood duck and are also quite popular in their natural colors. All of these feathers feature barred markings.

***Wing Quill*** These are the flight feathers from the wings of larger game birds. Again, waterfowl predominate. You'll hear the terms *primaries, secondaries,* and *tertials.* These refer to the portion of wing where a particular feather is found. Wing quills are a traditional—in fact, historical—material. They're mainly used to wing wet flies, but can be used for dry-fly wings as well.

***Hungarian Partridge*** This is a game bird that was imported from—you guessed it—Hungary. Of all the species that have feathers suitable for this sort of tying, the Hun is the best. A pelt holds hundreds of beautifully marked little feathers that are perfect for wet-fly hackles and nymph legs. As of this writing (December 1995), Hun pelts sell for around $20. Their utility is extensive, and in this price range they represent a real bargain.

***Beard*** This is a common form of wet-fly hackle—usually a small bunch of feather barbs tied in at the throat.

***Collar*** Hackle that is wrapped in such a manner as to form a cornucopia around the front of a fly is called collar. Collar hackle can also be stroked downward to form a beard-type hackle; in fly-tying parlance, this is known as a *true hackle.* It produces a fuller effect and is favored by salmon-fly tyers.

***Folding a Feather*** This is an important technique, but not a difficult one. It involves stroking the barbs to one side of the quill, as will be shown and explained in the tying exercise that follows. I strongly suggest that you practice this maneuver until you've mastered it, because it has many important applications in fly-tying, as you'll see.

### *About Wet-Fly Hooks*

The longer-shanked hooks you used for the Woolly Worm and Bugger are wet-fly hooks, but rather long ones. Wet-fly hooks are most commonly distinguished from dry-fly hooks by their weight, or wire thickness. Some manufacturers use a slightly different metallurgy for wet-fly hooks, making them a bit softer than the high-carbon dry-fly models.

Here, you'll be using models with a much shorter shank. They're generally labeled "Standard Wet Fly," or perhaps "1X-Long Wet Fly." As I mentioned previously, there are inaccuracies and ambiguities in the "X" method of rating hooks, but at least for the time being, we're going to have to live with it.

I prefer to think of a "standard" shank length as being twice that of the gape of the hook. The dry-fly hooks you'll be using farther on are such. For the wet flies that follow, you'll use hooks that are described as both "Standard" and "1X-Long,"—or "1XL," as the package usually says.

On that note, let's tie the simplest of wet flies, the soft-hackle. There's an enormous family of such flies, made with a wide range of materials and unique designs. The one thing they all have in common is that they're wingless. They've been around, in one form or another, almost since the dawn of fly fishing and are justifiably popular to this day. They have a nice action and appearance in the water and do a fine job of simulating the emergent forms of aquatic insect life.

**Thread Note:** I believe in using thread of a color that's compatible with the colors of the fly. In the case of subsurface flies, this is really quite important, because when they get wet the thread color becomes a factor in their overall coloration. If you were to use, say, a black thread with a light-colored body material, you'd notice a distinct discoloration when the fly got wet.

The rule of thumb is this: If you don't have exactly the color of thread called for in a given dressing, err on the light side. For example, brown thread is specified for the Hare's Ear to follow for two reasons: It's compatible with the overall color scheme of the fly, and it is the right color for the head. If you don't happen to have brown thread handy, but do have beige, or light gray, or tan, or even white, here's an option: Use that thread, then (before applying head lacquer) tint the head with a brown waterproof marker. You can either rub a few inches of thread with the marker before creating the whip-finish, or simply color the head afterward. Just be careful not to discolor the fly in the process.

The soft-hackle family of wet flies is quite historical and includes many dressings, both old and new. Soft-hackle wet flies are wingless and feature a collar-style hackle that envelops the body in a sort of cornucopia fashion. The pattern that follows can be tied both plain and with gold ribbing. I've included instructions for the ribbing and will leave it up to you as to whether you want to try it your first time through.

***The Dressing*** Hare's Ear Soft-Hackle

*Hook:* Daiichi #1550; standard wet-fly.

*Hook Size:* For this exercise, size 10 or 12.

*Thread:* Fine; 8/0 Uni-Thread or comparable; brown.

*Body:* Dubbing of the hare's-ear type.

*Hackle:* Hungarian partridge or similar feather.

### Tying Steps

1. Tie on near the front of the hook, and wrap the thread to the bend.

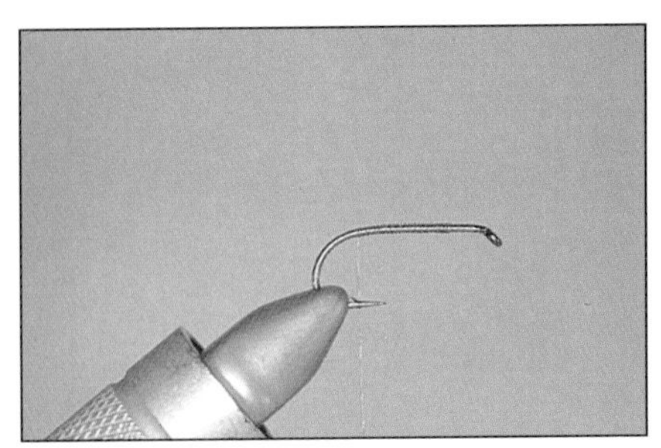

***Thread attached; step 1.***

2. You're now going to execute the *spinning-loop* method of applying dubbing, which I prefer for coarser materials. This involves locking the dubbing between two threads, twisting everything into a "rope," and wrapping it around the hook. Have a small quantity of hare's-ear dubbing at hand.

3. Run 5 or 6 inches of thread out of the bobbin. If you feel that a little wax would be helpful, apply some to the thread now. With your right thumb and forefinger, begin to spin very small wisps of dubbing onto the thread. The preferred method is to spin clockwise, with your thumb moving toward the left, because this twists the dubbing in the same direction that the thread will eventually be wrapped, which helps prevent loosening. The keys—and this applies to all dubbing—are to tease it out in wisps and to put on just a little at a time. Try to keep it even along the thread, with no lumps or bald spots. For a size 10 hook of the type specified, you'll need 2 to 2¼ inches of dubbing; for a 12, a little less. It's important, when following the spinning-loop method, to use just the right amount of dubbing; it isn't easy to add or subtract, as it is with the single-thread method. Also, you want to use up the dubbing just as you complete the body, so you won't have any excess to tie down; that creates a mess.

4. At this point, inspect your dubbing. If there's a length of bare thread between the hook shank and the spot where the dubbing begins, try to slide the material up toward the hook. Take your hackle pliers and grip the thread about ¼ inch below where the dubbing ends; then, while holding the pliers under tension with your left hand, pass the thread up and over the hook at the same point as the thread. This forms the dubbing loop.

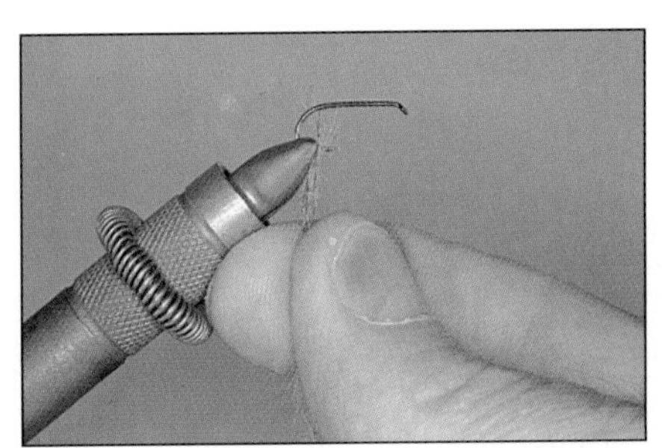

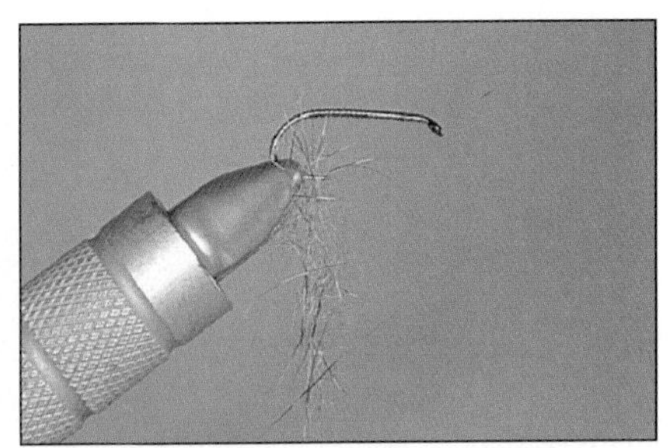

***Applying the dubbing; steps 2-4.***

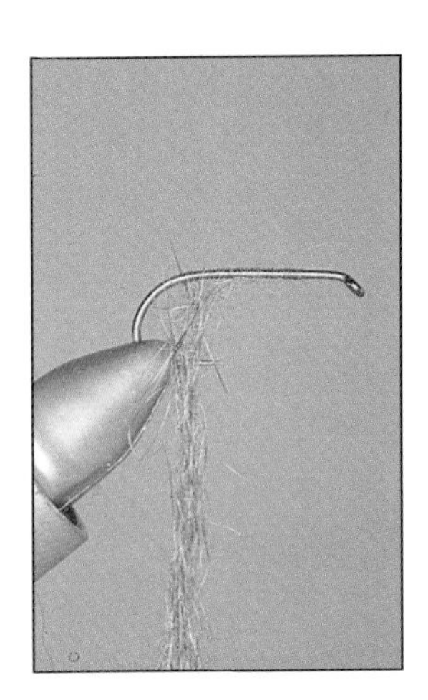

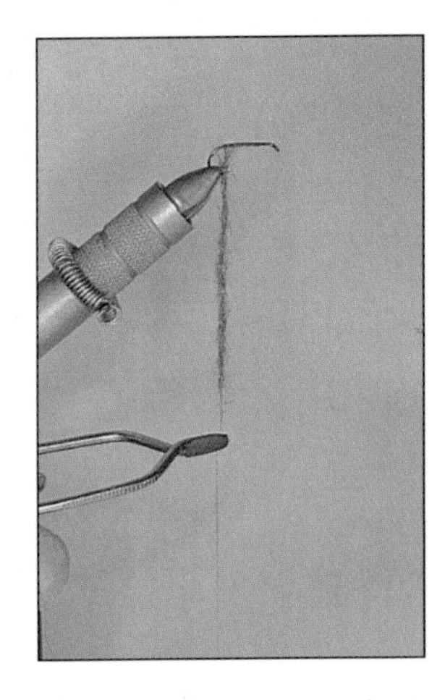

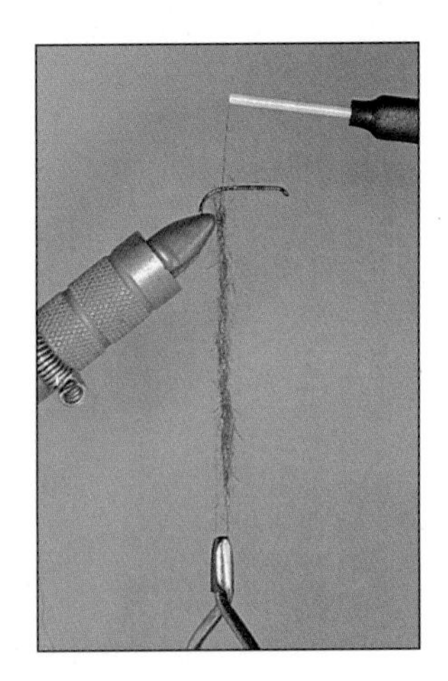

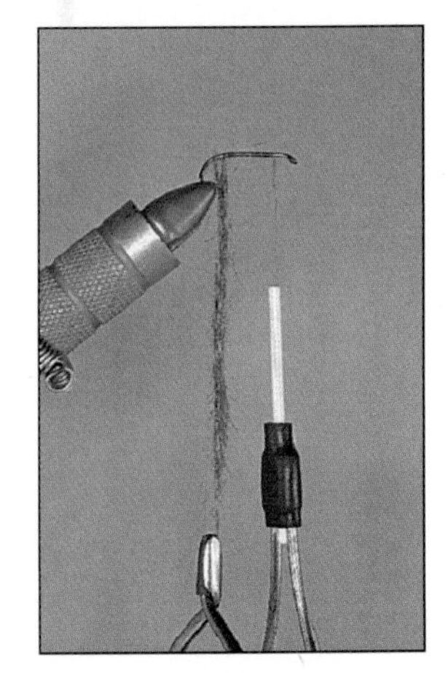

***Forming the spinning loop; steps 4-5.***

5. While maintaining tension with the hackle pliers, wrap the thread forward to where you want the body to end. You'll need enough space for a couple turns of hackle

and a whip-finish, but not quite as much as you'd need on a winged wet fly. Let the bobbin hang.

6. Grab the hackle pliers by the jaws and give them a good clockwise spin. You'll see the dubbing form into a tightly packed worm, or rope. Spin until the dubbing resists being spun further.
7. Using the hackle pliers, wrap the dubbing forward until it reaches the point at which you left the tying thread hanging. If you see that you need to tighten up at any point, just spin some more. Tie off the doubled thread that formed the spinning loop, and trim the tag end. If you find that you've put on too much dubbing after all, let the hackle pliers hang and allow the excess to unspin. Pick off as much of the dubbing as you can before tying off the thread. Then, wrap neatly to the eye and back to the front of the body, thus creating a thread base for the hackle.

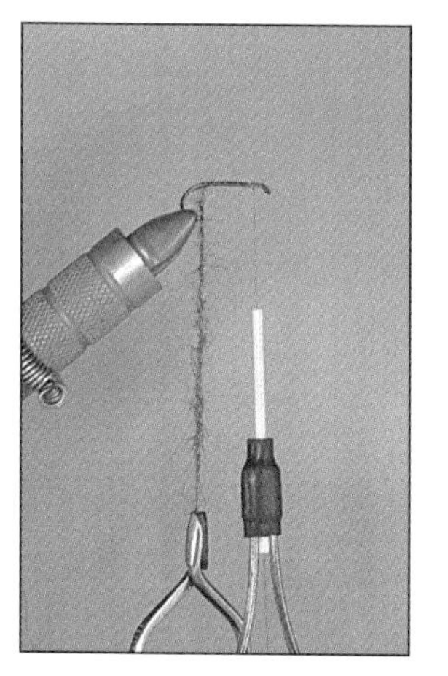

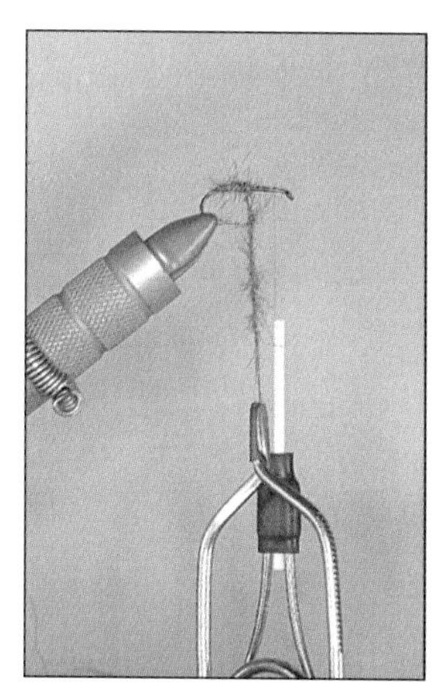

***Spinning and wrapping the dubbing; steps 6-7.***

8. Select a small mottled feather from the saddle of a Hungarian partridge. Gently, to avoid breaking the fragile quill, pull off the waste material near the butt end; then hold the feather by the tip and gently stroke the barbs to a 90-degree angle to the quill. At this point, the feather should resemble the one in the photograph.
9. Tie in the feather immediately ahead of the body by the tip end, as shown in the photo. Secure it well, so that it won't slip out while being wrapped. Trim off the surplus feather, and take a few housekeeping wraps to neaten and smooth the area.

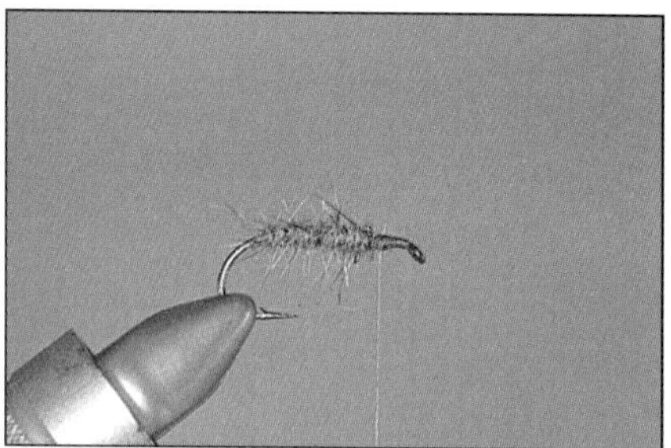

***The finished body, before and after trimming; step 7.***

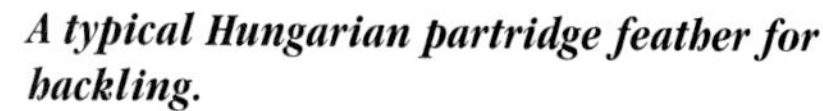

***A typical Hungarian partridge feather for hackling.***

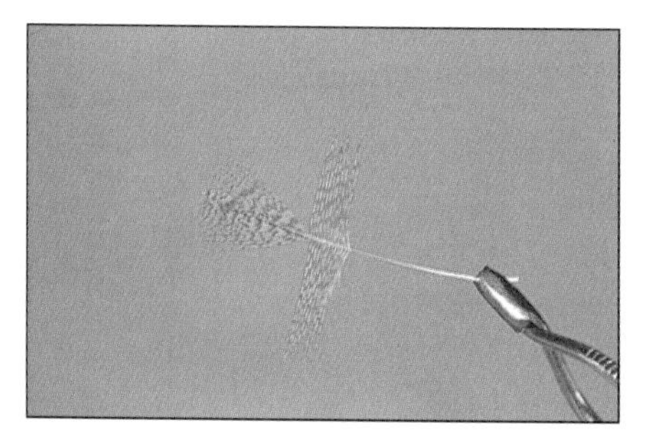

***Preparing and tying in the feather; steps 8-9***

10. When wrapping the hackle, you may or may not wish to use your hackle pliers. Hungarian partridge quills tend to be rather fragile, and I've found that I break

them less often if I wrap by hand. However, the choice is yours; the result will be the same. Here, we'll use them. As you wrap, gently stroke the barbs to the rear—folding the feather, so to speak. The quill is the leading edge, with the barbs trailing behind. The hackle should form an attractive cornucopia around the front of the fly. You'll only realize about two turns from a Hun feather, which is ideal, because that's all you want. This is very webby material, and too much hackle clutters up the fly.

11. When tying off this kind of feather, it works best to hold the quill to the side with your hackle pliers; this allows you to work in the thread between the fibers and catch the quill. Bind it down securely with four or five firm wraps, then trim off the tag end. Take a few housekeeping wraps, whip-finish, cut off the thread, and apply head lacquer.

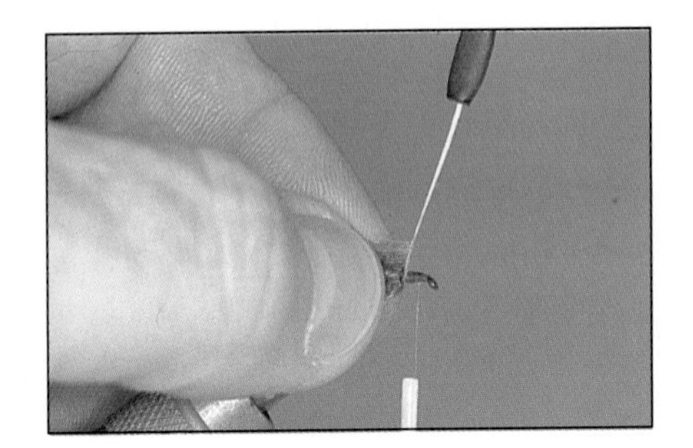

*Wrapping the feather; step 10.*

*Trimming the excess quill; step 11.*

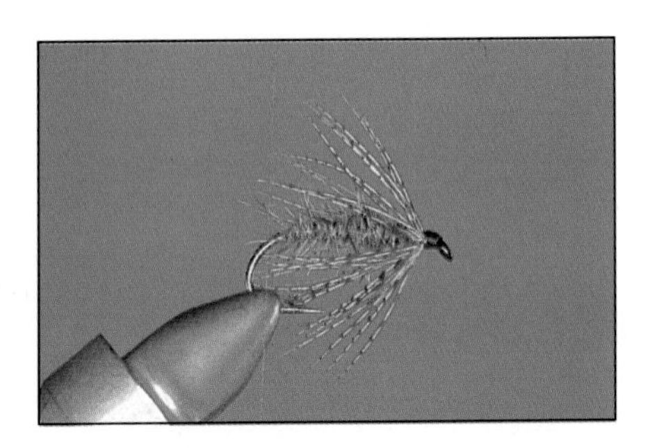

*The completed Hare's Ear Soft-Hackle.*

Now let's tie a winged wet fly and learn something about tailing, single-thread dubbing, wet-fly hackle, and, of course, the traditional wet-fly wing.

## Tying Term

*Two opposing wing quill feathers, with slips for wings removed.*

***Barb-Gathering*** This procedure is virtually the same as folding a feather, except that the barbs are removed from the quill by either cutting or pulling. It's one of the most important techniques in all of fly tying, and I encourage you to master it. You'll be using this procedure to gather tailing material, bunches of barbs for beards, legs for nymphs, and such. Ideally, you'll take barbs from both sides of the quill at once. However, for this to produce the desired result, the feather you're using must be symmetrical: The barbs on either side of the quill must be equal in length, or nearly so. If they're not, then you must gather the barbs from one side at a time.

### *About Wing Quills*

As I mentioned previously, these are the flight feathers from the wings of game birds. Feathers from left and right wings have opposing curvature; both sides are needed for tying a set of wings. Many species of waterfowl have the type of feathers required for winging wet flies. My favorite is Canada goose; however, smaller flies are better accommodated by the finer quills from duck wings.

The more similar the shapes, sizes, and textures of the two feathers, the easier it is to tie great-looking wings. It is *very* important to use the softer portions, avoiding the coarse material nearer the tips. This is particularly true in the case of primary, or leading-edge, flight feathers; however, they may have some nice stuff nearer their butt ends. Overall, the secondary feathers, which are the third or fourth ones back, are the best.

A slight amount of curvature is desirable, as it helps hold the wings in position, but too much will cause problems. While it's possible to remove some curvature by steaming, or ironing with a steam iron, it's best to avoid feathers with extreme curvature when you're purchasing.

***The Dressing*** Iron Blue Dun

*Hook:* Daiichi #1560; 1X-long wet-fly.

*Hook Size:* For this exercise, size 10.

*Thread:* 8/0 Uni-Thread or comparable; black.

*Tail:* Dark gray hackle barbs.

*Body:* Dark gray dubbing, natural or synthetic.

*Hackle:* Soft, dark gray hackle barbs, tied beard-style.

*Wing:* Two sections from opposing (left and right) gray duck or goose flight quills.

### Tying Steps

1. Tie on a short distance behind the eye, and wrap to the bend.

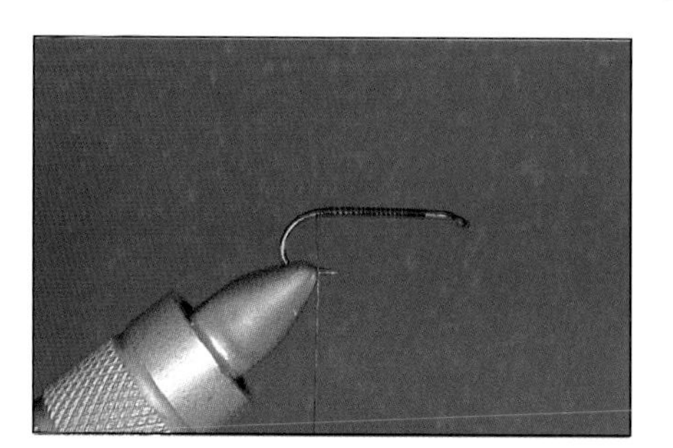

***Thread attached; step 1.***

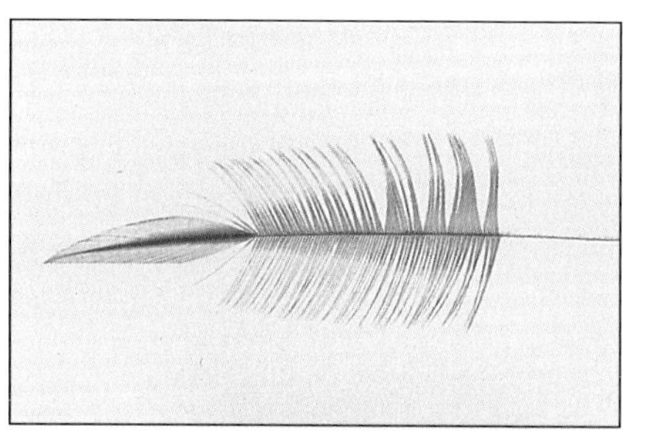

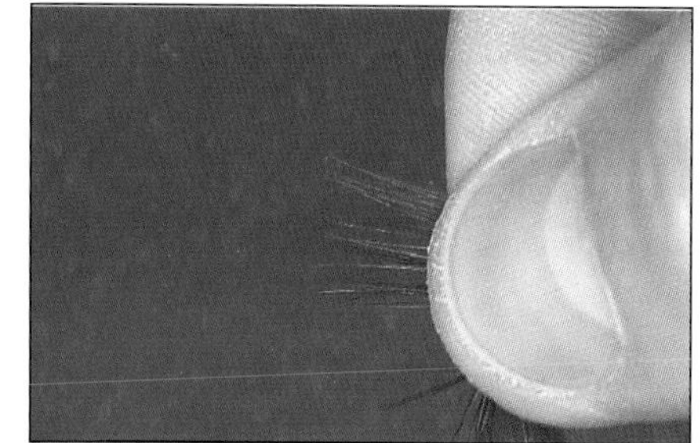

***Feather for tailing, and barbs gathered; step 2.***

2. Select a large gray hackle feather, hold it by the tip, and gently stroke the barbs outward to a 90-degree angle from the quill. Tear or cut off a small bunch from each side; these will compose the tail. Or, if you prefer, look ahead to the Dark Cahill tying sequence, study the "gathering barbs" procedure, and treat this feather in the same manner. Either way, try to keep the tips of the barbs neat and even.
3. Hold the bunch with your right thumb and forefinger, and gauge the length. The tail should be no longer than the hook shank, and perhaps a bit shorter, depending on the hook you're using.
4. Set the tail in position at the top rear of the hook, and transfer the tailing bunch to your left hand. Tie it on with several pinch-wraps, then check to see that it hasn't rolled to the far side. If it has, bring it back to the center; it should respond to being moved back on top beneath the thread wraps.

5. Secure the tail by binding down the butts, wrapping forward with moderate tension to a point about $^3/_4$ of the way up the hook shank. If the butts are so long that they must be trimmed, cut them on a slope, so that a tapered underbody will be formed.

6. Now you have a choice: You can use the spinning-loop method to construct the body, or you can simply spin on the dubbing and wrap it. Here we'll use the latter technique. Expose about 3 inches (75mm) of thread. If you plan to use additional wax, rub it onto the thread now. Then, spin on the dubbing, tiny bits at a time, teasing it off the main bunch in wisps so that it doesn't bunch up. Try to lay the fibers lengthwise along the thread. Apply plenty of thumb/finger pressure while you're spinning, as this will cause a little heat to build up, which will activate the wax on the thread and aid the process. For a size 10 hook of the type specified, you'll need 2 to $2^1/_2$ inches (50–62mm) of material on the thread, depending on how thickly you apply it.

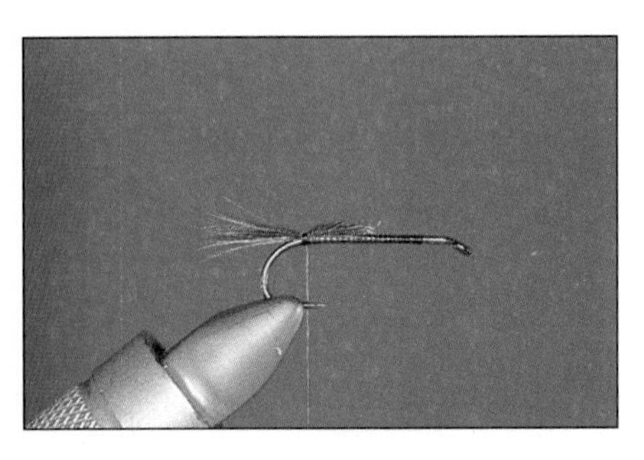

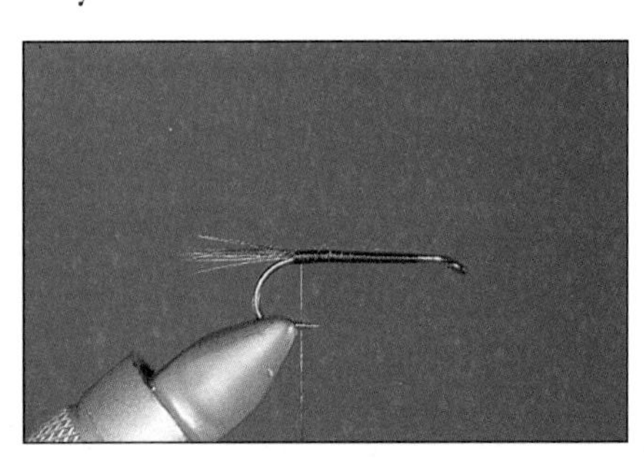

*Mounting tail and binding down butts; steps 3-5.*

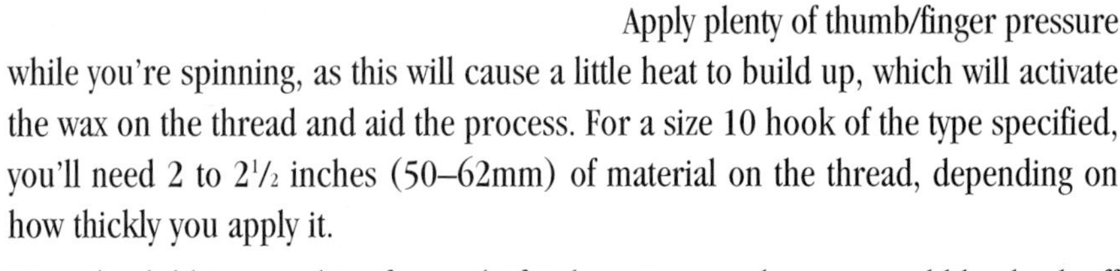

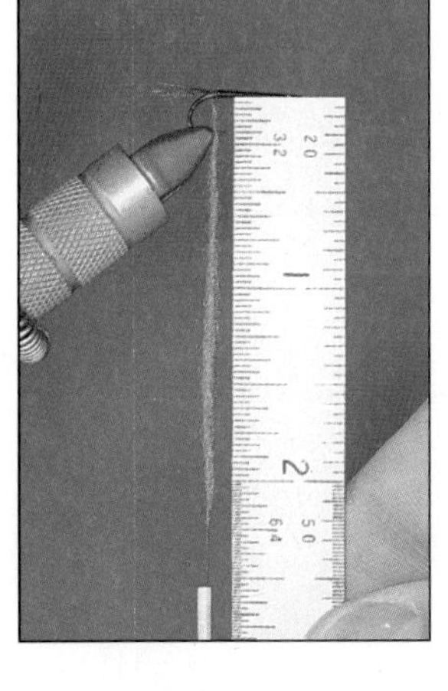

*Dubbing spun on for single-thread operation; step 6.*

7. Wrap the dubbing, working forward. If it doesn't go on the way you'd like, back off and reconfigure, adding or removing material where necessary and respinning. When you reach a point about 15 percent of the shank length from the eye, stop. Remember that there are still wings and hackle to be added, so don't crowd the eye.

8. Select a dark gray feather from either a hen or rooster cape, or a hen saddle—one that has fairly soft barbs. Check to see that the barbs are of equal length on each side of the quill. If they are, prepare it as described under step 2, the prefolding process, but more so. Using your right thumb and forefinger, stroke until all the barbs are on one side of, and at 90 degrees to, the quill. When you see that the tips are even, grab them with your left thumb and forefinger, and either cut or pull them from the quill. In either case, hold on tight, so that the tips stay even. If the barbs are not of the same length on both sides of the quill, simply use barbs from one side at a time to ensure that the tips are even.

9. Pass the bundle back to your right hand, holding it by the butts. Bring it up underneath the throat of the fly, right where the body ends. Gauge the length: The tips of the barbs should extend just to somewhere between the point and the barb of the hook. You have some discretionary leeway here.

   Tie the bunch in place with several upside-down pinch-wraps, then secure it with several more firm wraps. Trim off the excess butt material, and take a series of housekeeping

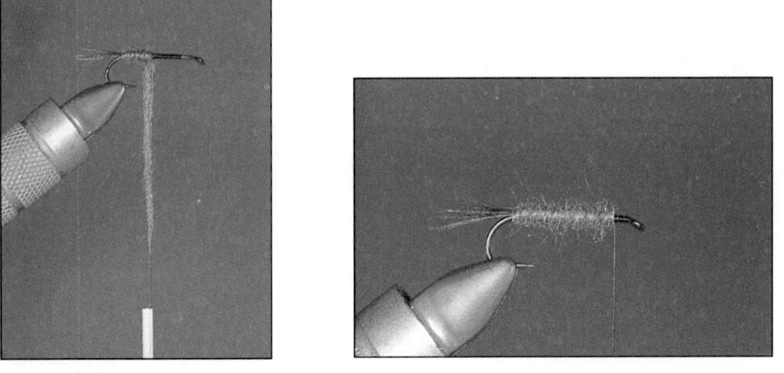

*Wrapping the dubbing; step 7.*

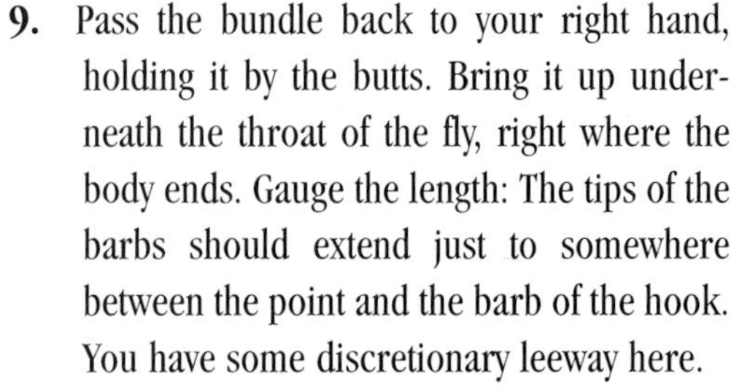

wraps to the eye and back to establish a thread base for the wing. *It's very important to have a full, smooth thread base!* The thread position preparatory to tying on the wings is just a turn or two ahead of the rear of the thread base. Observe the photo closely.

10. The wings are mounted precisely at the tie-off point of the hackle, so that there's no gap between the two. The first step is to remove two sections—or *slips,* as some tyers call them—from opposing wing quills. Size, of course, is relative; for this fly, on a size 10 hook, the slips should be about 3/16 inch (5mm) in width, or a bit less, and absolutely equal.

11. Hold the two slips by the butt ends in a sort of chopsticks fashion, concave sides facing, and manipulate them to get the tip ends absolutely matched up. Secure them by grabbing them for just a moment with your left thumb and forefinger, then passing them back to your right hand.

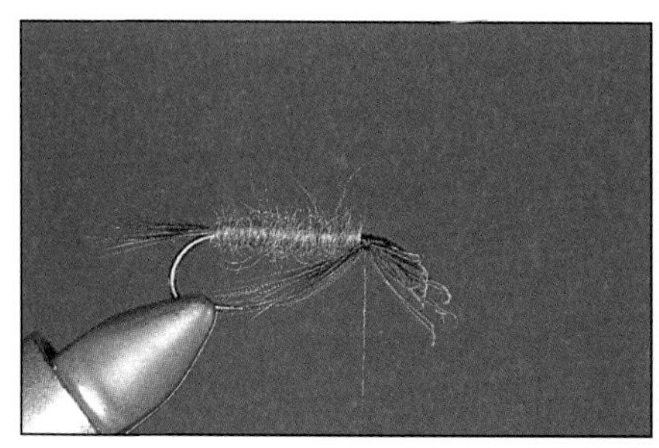

***Beard hackle in place, butts trimmed, thread base for wings established; steps 8-9.***

12. Set the wings precisely on top of the hook shank. They should be centered and exactly in alignment, front to back. Look at them from the top as well as from the sides. At this point, I'll mention that your tying position has a lot to do with your getting this type of wing on straight. If you sit too low in your chair compared to the level of the hook, your tendency will be to set the wings tipped at an angle toward yourself. The tips of the wings should rather extend rearward to a point where they would touch, or perhaps go very slightly beyond, an imaginary vertical line tangent to the rear extremity of the hook bend. Please don't make them too long; that's poor form.

13. Tie the wings in place with a series of carefully executed pinch-wraps, all the while holding them in position. What's happening—but of course can't be seen—is this: As the first pinch-wrap is tightened, the loop of thread compresses the wings both from above and from the sides. This causes them to slip *very* slightly down around the hook on their respective sides. However, don't *purposely* shroud the hook with the wings; let it happen naturally as a function of the pinch-wrap. The far wing will have a tendency to roll downward, due to thread torque. Counteract this by using your left forefinger as a "dam," if you will. Refer to the illustrations on page 42.

***Matching up the winging slips; step 10.***

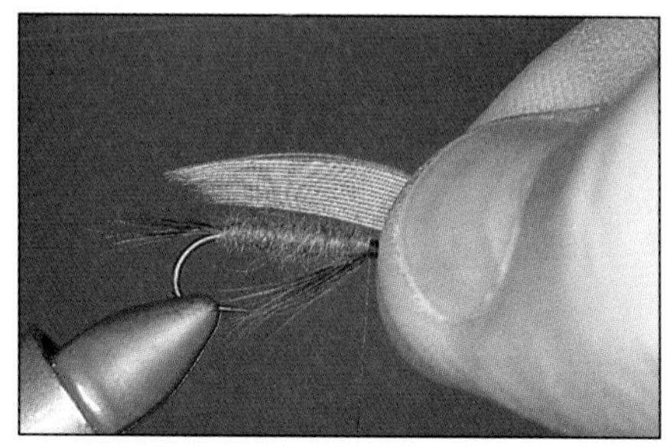

***Positioning the wings, gauging length; steps 11-12.***

14. Inspect the wings. If correction is required, gently restore them to their proper position.

**15.** Now seize the wings again with your left hand and take two or three more pinch-wraps, followed by several firm regular wraps. End with the thread hanging from the rear of the head, as shown. Now let go of the wings, and make one last check of their position before you trim the butts. Keep in mind that as long as the butts are attached, they are your allies: You can use them to adjust the wing position, if necessary. Once you start trimming, however, they become your enemies, because no matter how sharp your scissors or good your technique, the shearing action will seek to move the wings under the thread wraps. Guard against this by holding onto the wings while you're trimming and by doing the job in several small bites rather than one big chop.

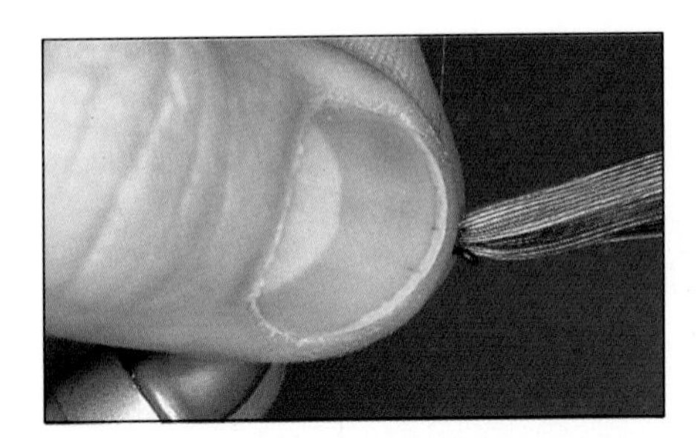

*Tying on the wings with the pinch method; steps 13-15.*

*The completed Iron Blue Dun.*

**16.** When your trimming is complete, take some housekeeping wraps to cover the trimmed ends, but be careful, as thread action can still upset what you've done. Hold onto the wings and don't apply a lot of tension, especially not when you're working at the very front of the trimmed butts. Start the whip-finish at the rear of the head and work forward.

Now that your fly is done, I'll reveal that you've just completed one of the most difficult and peril-fraught procedures in all of fly tying. The quill-section wet-fly wing has been giving tyers fits for generations, which I believe is the main reason that we don't see them as much today as in the past. But now you can create them and show them off to your friends—not to mention the trout.

Next, you'll tie a famous old Catskill wet fly, the Dark Cahill. In this exercise, you'll learn how to make a tail, a flank-feather wing, and rehearse the single-thread dubbed body.

I prefer a slightly longer hook for winged wet flies. Everything seems to fit better, and a prettier overall silhouette is achieved.

## Tying Term

***Soft Dubbing*** As differentiated from hare's-ear dubbing and such, this material is fine in texture—almost cottony. It packs well to form smooth, slender bodies on both wet and dry flies. It may be synthetic, natural, or a mixture of the two. With natural furs, the guard hair will have been removed, and only the soft underfur will remain. Beaver, muskrat, rabbit, fox, and certain other animals yield this type of underfur.

***The Dressing*** Dark Cahill Wet Fly

*Hook:* Daiichi #1560; 1X-long wet-fly.

*Hook Size:* For this exercise, size 10 or 12.

*Thread:* 8/0 Uni-Thread or comparable; black.

*Tail:* Brown rooster hackle barbs.

*Body:* Medium gray dubbing; soft-textured.

*Beard:* Soft brown hackle barbs; same color as the tail.

*Wing:* Lemon-barred wood duck flank feather or dyed substitute.

## Tying Steps

1. Mount the hook, tie on near the front, and wrap neatly to the bend.
2. Select a large hackle feather from the edge of a brown cape or neck. It should have long barbs, but they don't have to be stiff; save that kind for your dry flies. Fold the feather exactly the same as you did with the partridge feather for the Hare's Ear.
3. Gather the barbs from both sides, so that the tips are brought into alignment under the control of your right thumb and forefinger. Then transfer them to your left hand, grabbing them by the tips. Take your scissors and cut or tear off a small bunch of barbs from the base, where they meet the quill. Regrip them by the butts with your right hand.
4. You now have a small bundle of barbs, which you'll make into the tail. First, gauge the length: It should be that of the hook shank, or perhaps a bit less. Then position the barbs atop the hook, with the established length extending straight out to the rear, as shown.
5. Tie the bunch precisely atop the hook. You can use a couple of soft-wraps, if the barbs are stiff enough to accommodate this; otherwise, resort to the pinch-wrap, as described in chapter 2.
6. Ideally, the butt ends of the tailing barbs will be long enough that they can be bound down to form a smooth underbody. For best results, cut them on a slant, so that a gentle forward taper is formed. Then wrap neatly forward to the point to which you'll want the body to reach—about 80 percent of the hook shank. Positioning is critical here, so observe the photo closely. Then wrap rearward again to the bend. I should mention that when you're doing spinning-loop dubbing, advancing the thread forward after the loop is formed takes care of binding down the tail butts and creating a base for the body.
7. Now for the dubbing, which can be either spinning-loop or single-thread. These methods have been described in the two preceding tying exercises. It doesn't matter if the material is natural or synthetic—just so it's fairly soft and fine. The one thing to keep in mind is that the spinning-loop process will require a thicker worm

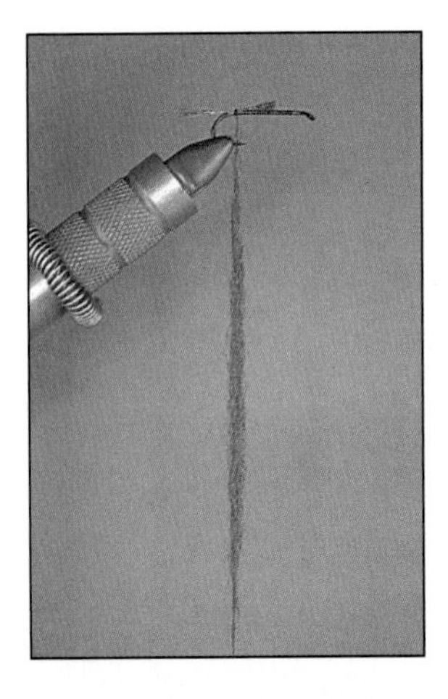

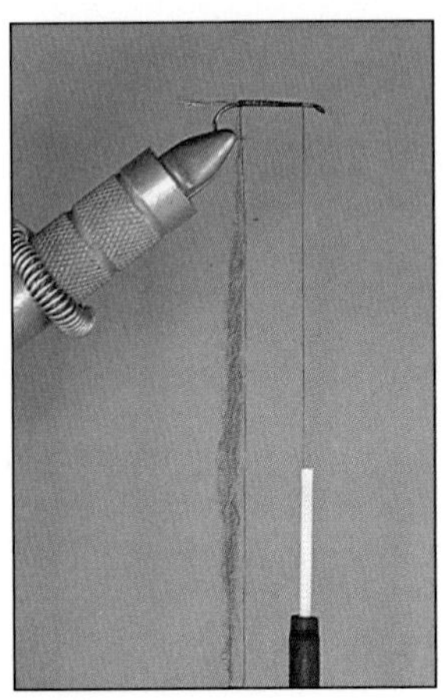

*Tail in place, dubbing spun on, spinning loop formed; steps 1-7.*

of material on the thread, because it packs tighter. Expose about 3 inches (75mm) of thread and spin on the dubbing, as shown.

**8.** Wrap the dubbing, working forward. If it doesn't go on the way you'd like, back off and reconfigure, adding or removing material as necessary and respinning. When you reach a point about 15 percent of the shank length from the eye, stop. Keep in mind that you'll need a little more working space up front on this fly than you did on the soft-hackle, because both hackle and wings are yet to be added. Refer to the photo for proportions. Create a thread base by running the thread neatly out to the eye and back.

**9.** Now for the beard-style hackle, which consists of a bunch of barbs similar to those used for the tail, but softer. You may be able to find these in the lower portion of a rooster hackle, or you may need hen hackle, which is invariably soft and webby. Gather the barbs as you learned in previous operations, procuring a fairly generous bunch. Cut or strip them from the quill, then pass them to your right hand, holding them by the butt ends with your thumb and forefinger.

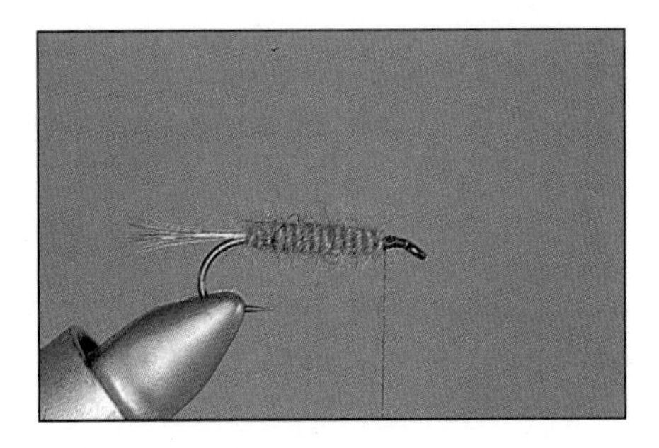

*Body wrapped, thread base established; step 8.*

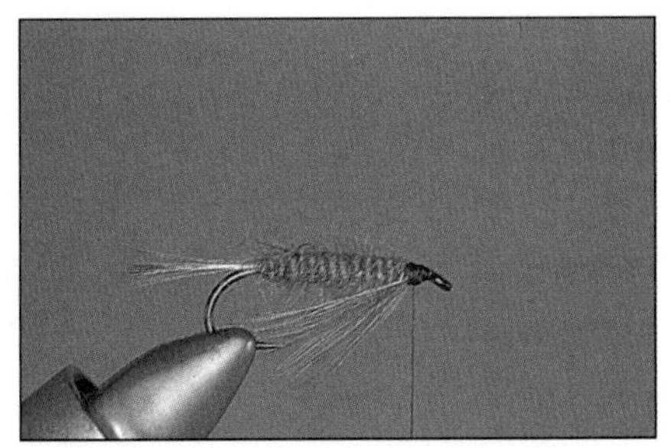

*Hackle in place, thread base for wing established; steps 9-10.*

**10.** Position the barbs beneath the throat of the fly. Gauge the length: In this instance, the tips should extend to about the point of the hook. Tie the bunch in place with several upside-down pinch-wraps and secure it with several more firm wraps. Trim off the excess butt material and take a few housekeeping wraps to establish a base for the wing. Note the thread position, which is just a turn or two from the rear of the thread base.

**11.** Dyed imitation wood duck flank works very well for this type of wing; here, we'll use dyed mallard. Select a good-sized feather with a centered quill and a neat, even tip. Check to see how far out toward the end of the feather the quill extends. You'll be using about $^{5}/_{8}$ inch to form the wing, so if the quill extends out that far, go in with the tips of your scissors and snip out just the quill. You'll lose a little material in the process, but not enough to matter. The important thing is that the quill, which contributes unwanted bulk and stiffness, is eliminated.

**12.** Strip off the shorter fibers on each side, so that the feather resembles the one in the photo. Then hold it in a flat position with your right thumb and forefinger, pretty-side-up. With your left thumb and forefinger, stroke downward over the fibers that protrude, folding them together into a neat bunch. You may find it helpful to moisten your fingers slightly.

13. Set the bunch in position on top of the thread base. Gauge the length: The wing should extend a little beyond the rear of the body, but not as far as the rear extremity of the hook. Now, tie the wing in place with several firm pinch-wraps, followed by several more securing wraps. Check to see that it hasn't been carried to the far side by thread torque; the wing should be centered, and if it's not, make a correction.

***Tying the wing in place; steps 11-13.***

***The completed Dark Cahill.***

14. When the wing is positioned properly and you're sure it's secure, trim as follows: Hold the wing tightly with your left hand and trim the butts as closely as possible without cutting or abrading the thread. Remember, you don't have to chop off everything in one cut; you may want to make two or three passes.
15. Cover the trimmed butts with neat wraps, whip-finish, and lacquer.

One further note about beard-style hackle. In order for it to look good and fish well, it needs to billow out somewhat from the bottom of the hook. This billow is affected by the use of soft material, which crimps under thread pressure, and to a lesser extent by the influence of the front of the body. Compare your finished fly to the one in the photo.

Also, an observation on this type of wing: Flank feathers vary somewhat in shape. Some of them make a nicer-looking wing when tied on upside-down, after being bunched. Something to keep in mind.

We'll complete the section on wet flies with a time-honored, colorful pattern that definitely falls into the attractor category: the Professor.

*Attractor*—that's an interesting term, isn't it? It refers to those flies that don't specifically resemble a particular item of fish food. But to whom are they really attractive—the fish, or us? Let's face it: This is a totally homogenized expression. Can you imagine a big old trout scrutinizing some garish bit of fluff and saying to himself, "Wow, that's a great-looking attractor pattern; I'm going to eat it!" No; fish aren't connoisseurs, they're simply predators. If they go for a fly, it's because they're hungry, and they think it's something good to eat.

The interesting and, in fact, miraculous aspect of all this is that fish often *do* go for attractor patterns. Sometimes they work when the best hatch-matchers fail. Then there's the Atlantic salmon—for that matter, the Pacific strain as well. These fish aren't eating during their migratory excursions into fresh water, so why would they want to take a fly of any sort? Many theories have been offered, yet we still have nothing more than empirical knowledge. It's an eternal question, and frankly, I'm glad.

In tying the Professor, you'll learn about floss and flat tinsel and also get a chance to practice the flank-feather wing that you just tied on the Dark Cahill. I should mention that in larger sizes, this type of fly is often dressed using a "true" hackle, rather than a simple beard. This technique produces a gorgeous effect and is widely used on full-dress Atlantic salmon flies and such. If you'd like to try this method, look ahead to the Shushan Postmaster exercise in the chapter on streamers.

## *About Floss*

Floss has been with us, in one form or another, virtually since the beginning of fly tying as we know it. The original material, silk, is still available, though more or less relegated to the venue of the traditionalist, who ties classic patterns and is interested in authenticity. The rest of us use modern synthetic flosses made of such materials as rayon (my preference), nylon, and similar synthetics.

Today's flosses come in a wide array of colors and thicknesses, including "hot" and fluorescent colors. There's been a fair amount of hoopla about the fluorescents, but as yet no votes are in from the fish. The way to determine whether a material actually is fluorescent is to put it under ultraviolet light and see if it glows.

The thickness factor is of considerable importance and will, to a large degree, determine how well your flies come out. Today you can get just about any denier you want, including minifloss for tiny flies. One popular floss product comes four-stranded, and you simply use as many strands as the size of the fly dictates.

There are now quite a number of flosslike products available, including stretch nylons, polypropylenes, and other contemporary synthetics. Some of these actually surpass true floss in certain respects. We'll have a look at them in the streamer-fly chapter.

## *About Tinsel Ribbing*

In this pattern, you'll be introduced to the process of ribbing a body, something you'll quite likely repeat thousands of times over the course of your tying career. As I stated at the beginning of this chapter, the most popular type of tinsel in use today is made of Mylar. The flat type comes in four widths that I know of and is silver on one side and gold on the other. This covers most tying requirements. Copper has come into prominence in recent years; you'll be using it in the nymph chapter.

Mylar tinsel is very forgiving and easy to use, compared to the old metallic tinsels that cut and frayed Granddaddy's silk threads. The only drawback is that it's not the strongest material in the world, particularly not in the narrower diameters used for most trout flies. Therefore, you must be careful when handling flies that contain this material; you don't want to grab a Mylar-ribbed fly around the middle with a pair of pliers when removing it from the jaws of a fish.

There are also several recently introduced tinsels that represent a middle ground between the coarse traditional tinsels and Mylar. They're stronger than Mylar and better replicate the effect of the old stuff, without the tying problems. The Uni company that gives us Uni-Thread also puts out an excellent line of such products.

## *About Dyed Hackle*

The material used for the Professor's tail comes from utility chicken pelts, many of which are a by-product of the food industry. Due to complicated regulations in this country and negative economics, most of them are imported.

They come in several forms: capes (necks), saddles, and strung bundles. They're useful for making streamers, bass and panfish flies, components of salmon and steelhead flies, and, in the example at hand, tailing on a wet fly. Typically, the feathers are large and fairly soft and have long barbs. For what you're about to do, that's just what you want.

***The Dressing*** The Professor

*Hook:* Daiichi #1560; 1XL wet-fly.

*Hook Size:* For this exercise, size 10.

*Thread:* 8/0 Uni-Thread or comparable; any pale color and black.

*Tail:* Bright-red-dyed rooster hackle.

*Ribbing:* Fine gold Mylar or similar tinsel product.

*Body:* Yellow floss.

*Hackle:* Brown hen or soft rooster.

*Wing:* Barred mallard flank feather.

## Tying Steps

1. Tie on about 30 percent of the shank length rearward of the eye with five or six wraps, then trim off the thread tag. You'll be tying in the tail at this point. Trust me; this is a bit different than on previous flies.
2. Select a large red hackle feather from the edge of the cape or from a strung bundle, if that's what you happen to have. The longer the barbs, the better. Barb-gather a bunch, per previous instructions. Hold the barbs over the hook, and gauge proportions for the tail, allowing an appropriate amount—about equal to the shank length—to overhang the bend.
3. Tie in the bunch on top of the hook with a couple of pinch-wraps, then secure with a series of neat, firm wraps, working rearward. With your left hand, hold the bunch at a slight angle above the hook, while pulling from the rear. Wrap all the way to the bend. This method saves you two layers of thread; the only requirement is that the tailing barbs be of sufficient length to facilitate the process. Lacking that, follow the procedure for the Dark Cahill.
4. Come forward a couple of thread wraps and tie in the ribbing tinsel against the far side of the hook; later on, this will allow for one turn of tinsel behind the tie-in point of the tinsel. If you're using Mylar, make sure the gold side is facing out. You can probably sneak a soft wrap or two around the material to set it into place. However, if the material flops around and gives you trouble, try this technique:

   (A) Hold the tinsel tightly between your two thumbs and forefingers.

   (B) Place it under the hook, just ahead of the thread.

   (C) With your left middle finger, catch it against the back of the hook.

   (D) Pick up the bobbin and pass the thread over the material.
5. Wrap forward very neatly; when a floss body is called for, contiguous wraps form the best base. Bury the tag end of the tinsel, cutting it off when you've worked about 3/4 of the way up the shank. Stop here.

6. Cut off a piece of floss 6 or 7 inches in length. Floss bodies should be slim, so select your floss accordingly. Here, I'm using two skeins of the four-ply Danville rayon floss. Tie in the material at the position shown, using four or five very firm thread wraps. I suggest the two-handed technique described in the previous step; it works even better for floss than for tinsel. After the first two wraps, do this: Pull the floss gently rearward, so that about 1/8 inch of it sneaks under the thread; this ensures that when you begin wrapping the floss, the filaments will all be lined up neatly. Secure with several additional wraps. Don't be concerned that a little thread bulk builds up; you'll take care of that in a few moments.

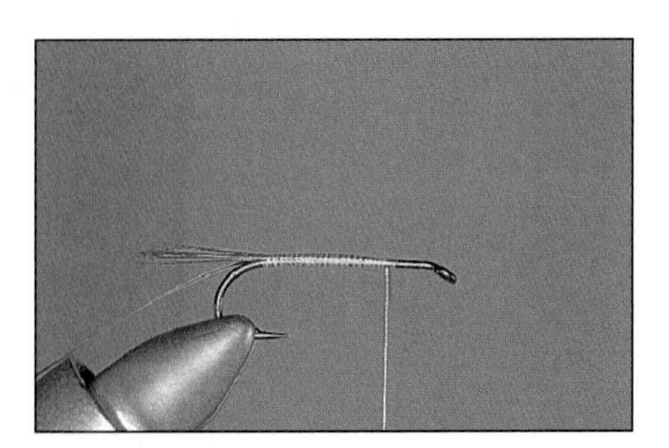

***Tail tied on, butts bound down, ribbing tied in; steps 1-5.***

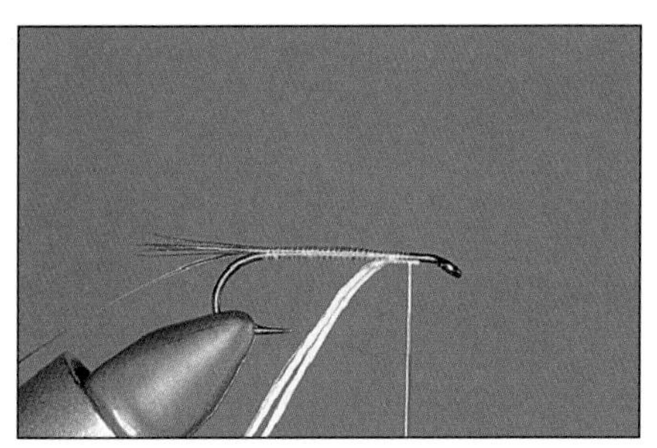

***Floss tied in; step 6.***

7. Before wrapping the floss, trim off the tag end. Then, wrap the material smoothly and evenly rearward, overlapping the turns a little. The floss will flatten as you wrap, but you don't want any filaments to separate from the skein. If you see this happening, twist the floss a little and keep on wrapping.
8. When you reach the bend, pass the floss once behind the tinsel; this sets it into position to start the ribbing. Then wrap forward, covering the first layer. You'll notice that the second layer goes on smoother than the first, which is why you use two layers.
9. As promised, you'll compensate for any bulk caused by the thread wraps used to tie in the floss. When you reach the tie-in point, simply back off all but one of those wraps. Complete the body with a couple more turns of floss, allowing it to push the thread along in front of it. Tie it off beneath the shank and trim the tag end. This should leave about 1/5 of the hook shank ahead for hackle and wings. This technique for reducing bulk can be used any time a material is double-wrapped. It's of particular value with flat tinsel bodies.

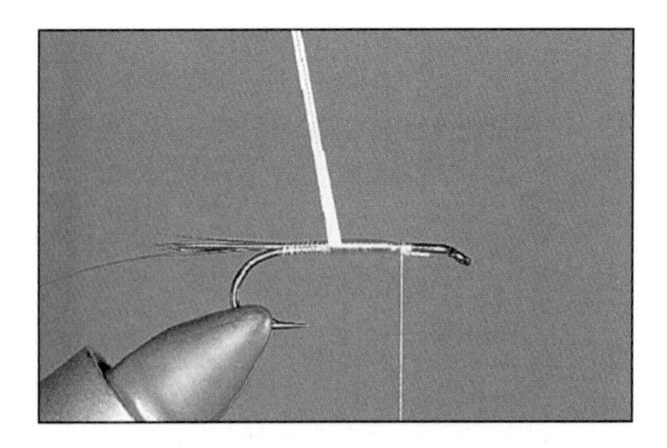

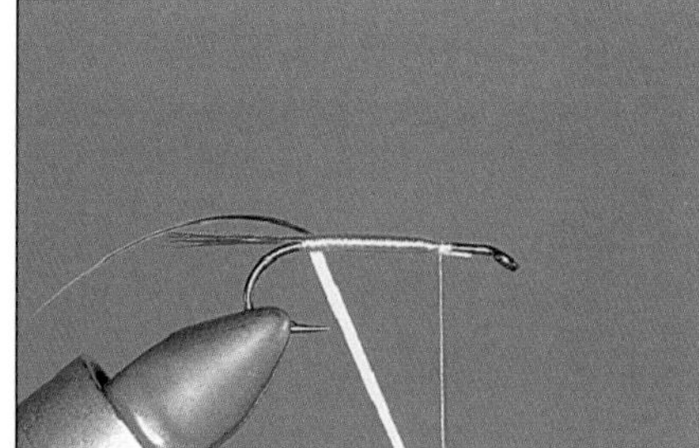

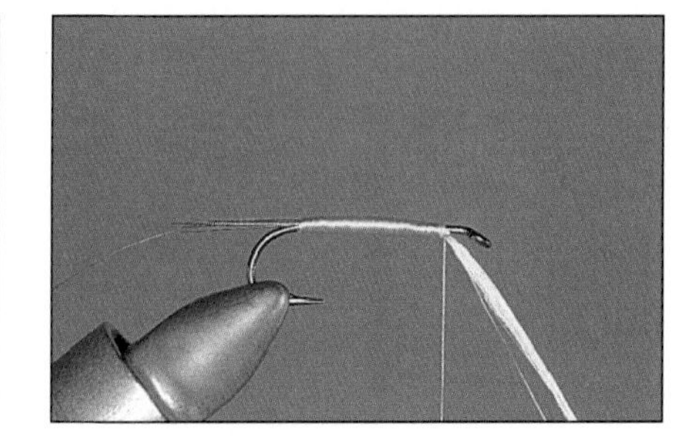

***Floss body being formed; steps 6-9.***

10. Now wrap the tinsel ribbing. Given the size of hook and width of tinsel involved here, you'll get five turns. Keep them evenly spaced, and try to come out in such a position that you can tie off the tinsel either on the bottom or against a side of the hook. Take a couple of housekeeping wraps.

**11.** The beard or throat hackle for this fly is exactly the same as for the Dark Cahill, in both color and construction. Please refer to that tying sequence.

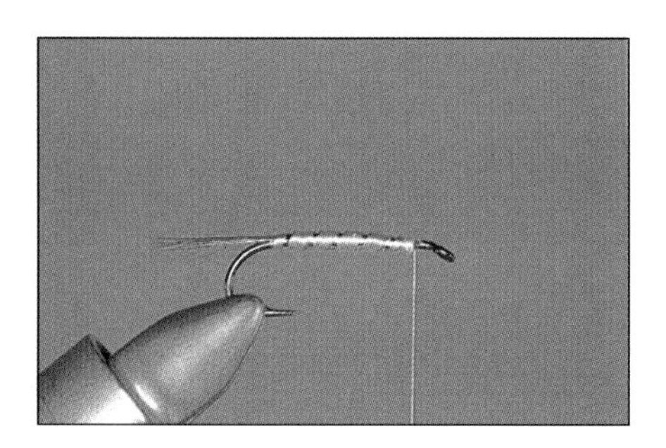

***Floss trimmed, ribbing in place; steps 9-10.***

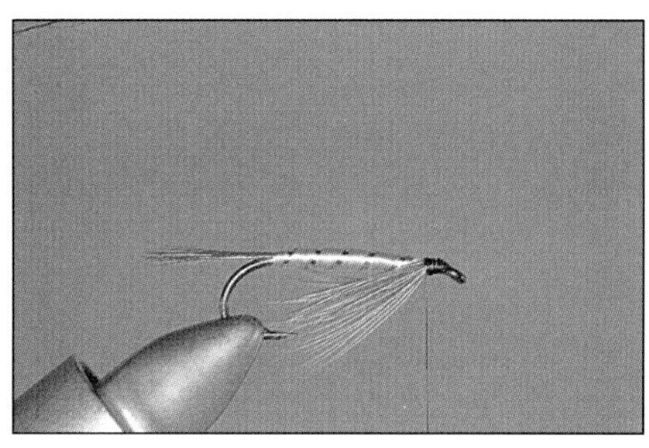

***Hackle in place, thread base established; step 11.***

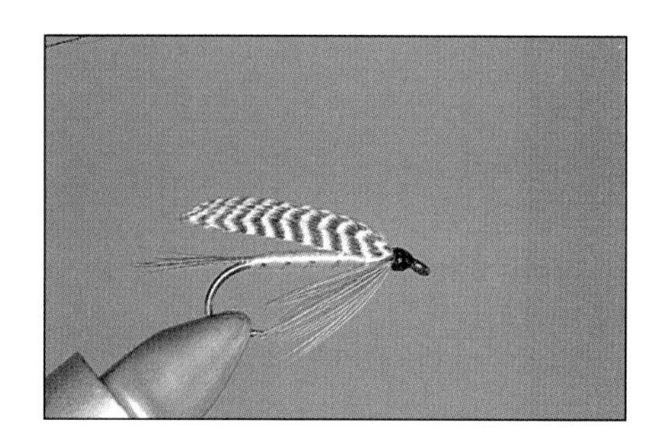

***The completed Professor.***

**12.** Switch to black thread and add a wing, exactly as you did on the Dark Cahill. Tie off, lacquer, and admire.

You can avoid changing thread color by two methods: either blacken the yellow thread with a felt marker before creating the whip-finish, or use black head lacquer. The latter is dangerous: one tiny misstep, and you'll have a messed-up fly. Any colored head cement or lacquer must be applied with great care.

The techniques learned from the four flies in this chapter not only enable you to dress a host of wet-fly patterns; they also constitute a valuable preparation for the other types of flies that lie ahead.

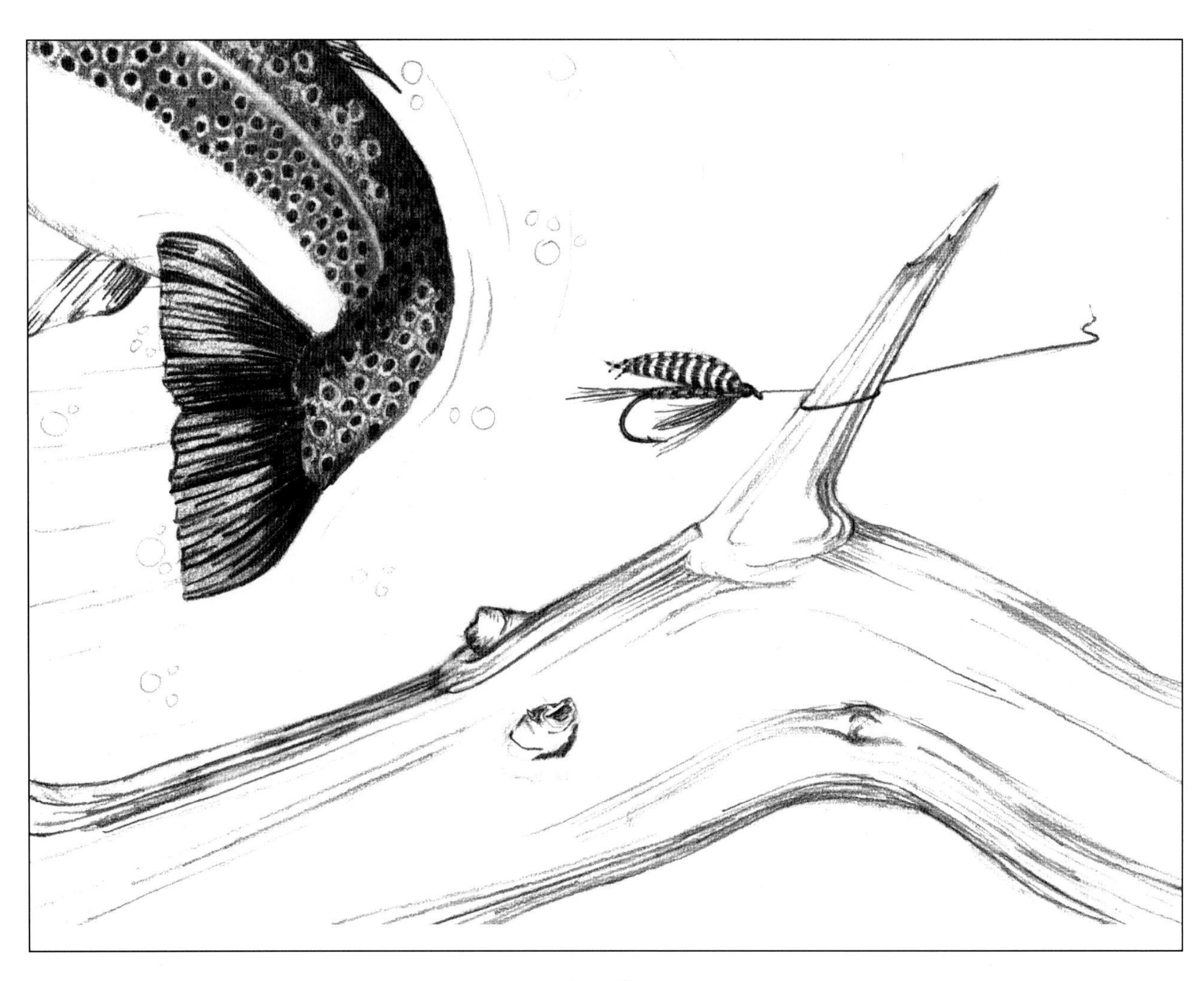

*A wet fly.*

◂ F O U R ▸

# *Nymphs*

In this chapter we'll work on what's become a highly popular category of flies: nymphs and related subsurface imitations. These are easy and fun to tie, and most gratifying to the beginner, as they're the easiest of all of the fly types to make look realistic.

As you're probably aware, nymphs are the subaquatic, immature forms of two very important orders of flies: the mayflies and the stone flies. Caddis flies, true midges, and certain other insects also have subaquatic stages that are very similar to, but technically not, nymphs. What you learn here will enable you to tie them as well.

Let's begin with a nymph design of fairly recent vintage. It's called the Flashback. This has nothing to do with the acid trips of the hippie era; the term refers to the tinsel or other shiny material that's used for the wing case—a "flashy back."

This particular design is quite generic in that many variations of it are tied, using different colors of dubbing, tailing, hackle, ribbing, and the flashy stuff itself. The one you're about to dress is a Gold Flashback, named for the color of the ribbing and wing case. It's a great favorite of mine and has really produced in every place I've used it, including New Zealand.

I think tying this fly will be a cinch for you, because it incorporates materials and techniques with which you became familiar in the wet-fly chapter. The only differences are the design and shape of the fly, particularly the wing case, which I believe you'll find easier to construct than the standard types of wet-fly wings.

I tie some of my nymphs with weight and some without. This one will be weighted, in a manner similar to the Woolly Bugger, but on a smaller scale.

## Tying Terms

***Mottled Hen Saddle*** These are feathers from the saddle of a female chicken. Some hens have beautifully marked feathers, similar to Hungarian partridge. They vary widely in shade and markings. The soft texture makes them ideal for nymph and wet-fly hackles and tails.

***Distribution Wrap*** In this variation of the soft-wrap, materials are caused to deploy around, or partially around, the hook.

***The Dressing*** Gold Flashback Nymph

*Hook:* Daiichi #1560 or #1710; 1XL or 2XL nymph, respectively.

*Hook Size:* For this exercise, size 10 or 12.

*Thread:* Dark gray, brown, or black.

*Tail:* Barbs from a mottled hen saddle or Hungarian partridge.

*Ribbing:* Fine gold-wire or oval tinsel.

*Underbody:* Fine lead wire; .010" diameter or at most, .015".

*Body/Thorax:* Any soft, dark dubbing.

*Wing Case:* Wide, flat gold Mylar tinsel or gold Bugskin (see the notes preceding the next pattern for a rundown on Bugskin).

*Legs:* Same as tail.

### Tying Steps

1. Tie on near the front of the hook and wrap back a little way. Now tie in the ribbing wire against the far side of the hook. Wrap the thread to the rear, binding down the wire. Bend it back and downward, so that it will not interfere with tying on the tail.
2. Tie on a short tail, just as you've done in previous exercises. In this case, after securing the tail, cut off the butts sheer rather than burying them with thread. This sets things up for the lead wire. Refer to the Woolly Bugger instructions.
3. Cut off a 6-inch piece of lead wire and wrap it around the hook in such a manner that it abuts the tail butts. There's no thread involved in this; the wire simply molds itself to the hook. Wrap it forward in contiguous turns to where the thorax will be located, and trim.

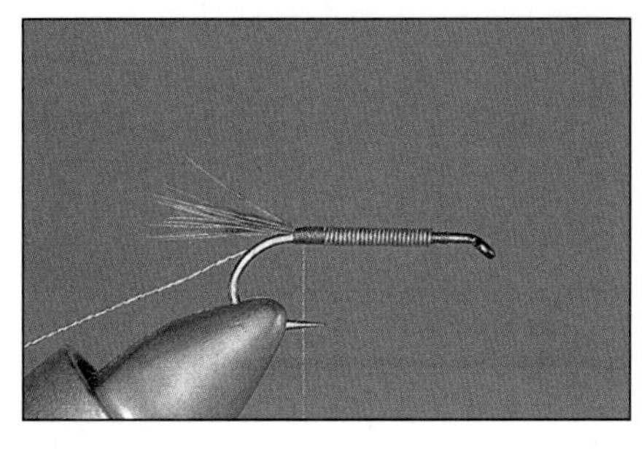

***Tail in place, ribbing tied in, wire in place. Note thread wraps forming a "ramp" at the rear of the wire; steps 1-4.***

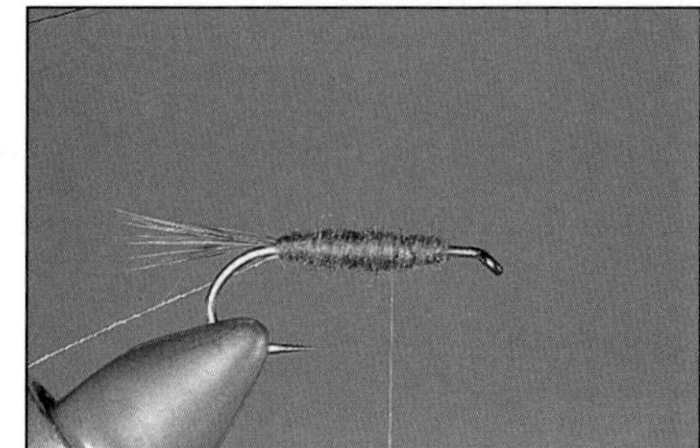

***Forming the body, positioning the thread; step 4.***

4. If there's a significant disparity between the tail butts and lead wire where they meet, take some thread wraps to smooth it out. Then make a dubbed body using the single-thread method. Make it fairly slender, and very gently taper it by gradually building

up a bit more dubbing as it's spun onto the thread. Stop the body just in front of the spot where the lead wire ends. Then wrap the thread back over the dubbing and wire a few turns. Proportions and positioning throughout this sequence are critical; study the photos closely.

5. Now it's time for the ribbing. With this fly, you have the choice of wrapping it in the same direction as the thread, or reverse-wrapping it—which I recommend when ribbing softer materials, such as dubbing and herl, with fine tinsel or wire. In either case, keep the turns well spaced and evenly spaced. Tie off the ribbing material in the thorax area. Don't worry that bulk is building up there; this is the widest part of a nymph, and you'll be covering up everything with dubbing in a moment.
6. Cut off a short piece of the flat gold Mylar tinsel—just long enough to be manageable—and tie it directly on top of the hook, in the position shown. If you're using the gold/silver double-sided type, tie it in with the silver side up, so that, when it's folded over to form the wing case, the gold side will be exposed. Secure it by running the thread forward, then back again, but don't wrap near the eye.

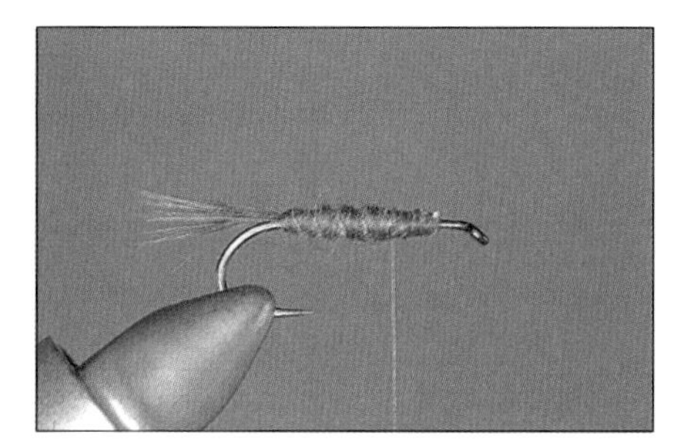

***Ribbing in place; step 5.***

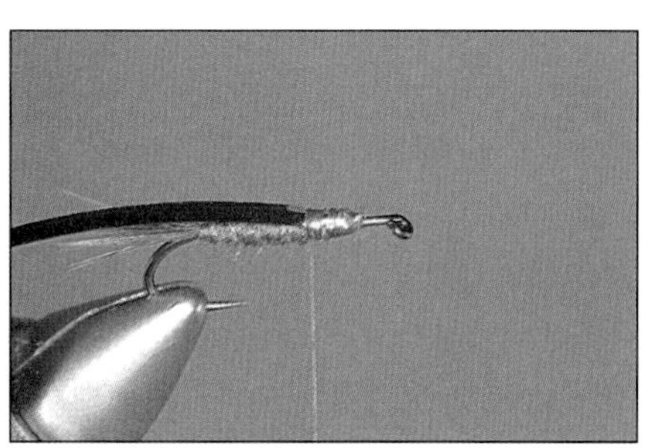

***Tinsel for wing case tied in; step 6.***

***Thorax dubbed in; step 7.***

7. Spin a little more dubbing onto the thread and wrap a thorax, covering the tie-in wraps used to secure the wing case material. Leave space up front for legs (hackle barbs) and for the tie-down of the wing case.
8. For the legs, take a small bunch of barbs from the type of feather prescribed in the dressing. Fan them out a little. With your right hand, set them on the far side of the hook, just in front of the thorax. They shouldn't be too long; the tips should fall a bit short of the point of the hook.

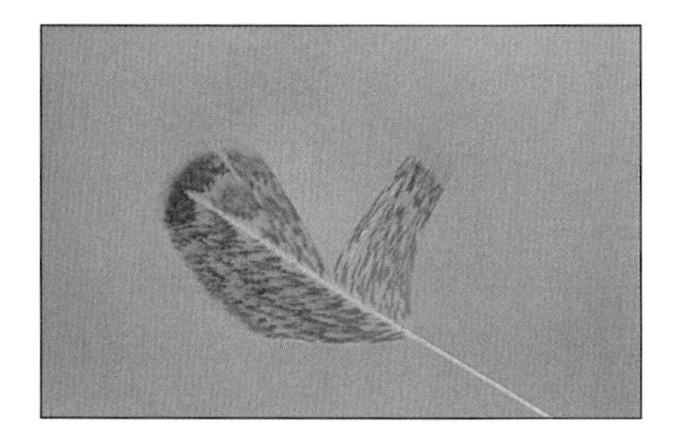

***A speckled hen feather, showing barbs for legs; step 8.***

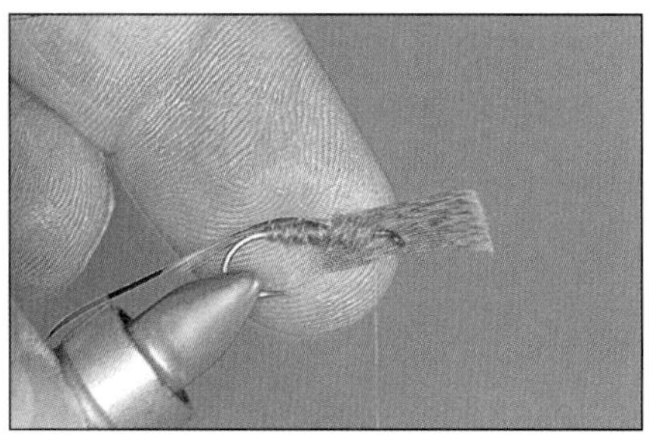

***Barbs in position for deployment; step 9.***

***Barbs deployed to form legs; steps 9-10.***

9. Now hold the bunch in place with pressure from the backside with your left forefinger. Soft-wrap the thread over the barbs, and use your left finger to cause the barbs to deploy around underneath the hook as the thread is wrapped, as well as up the near side a little. Gradually increase thread tension as you go. I call this a *distribution-wrap.*
10. Take a firm wrap or two, and inspect the deployment of the barbs. The idea is to end up with a reasonably even distribution of barbs about 180 degrees around the sides and bottom of the hook. You can adjust their position by rolling them back and forth a little with your right thumb and forefinger before immobilizing them with more wraps. When they're positioned as you wish, secure them with a few firm thread wraps and trim the butts.

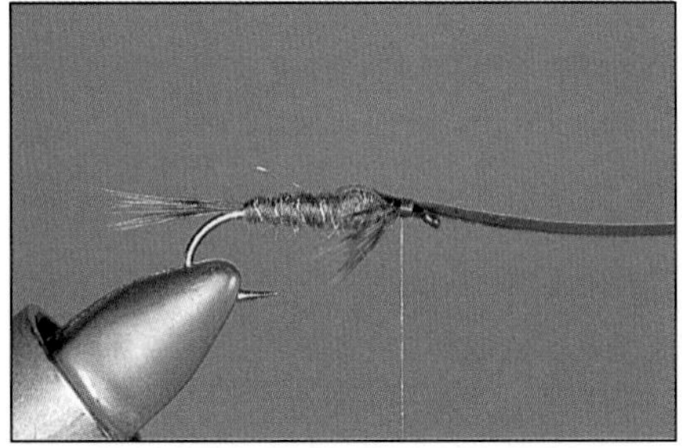

*Tinsel wing case folded forward and tied down; step 11.*

*Two views of the completed Flashback Nymph.*

11. Now fold the tinsel wing case forward over the top and tie it down with a series of very firm wraps. While it's not mandatory, I feel it's wise to fold the tinsel back over itself and take some more wraps over the doubled area. Tinsel is slippery, and this is insurance against it pulling out from beneath the thread wraps. Whip-finish, apply head lacquer, and take it fishing.

Flashbacks are also tied in other colors. I have a personal affinity for copper. The only problem I'm having is finding the right kind of material as, to my knowledge, there is no copper Mylar tinsel. The only flat tinsel I can find in copper is rather narrow and is a true metallic, which doesn't handle as nicely as Mylar. When tying with it, you must be careful: The edges are sharp and will cut thread. For larger flies, I tie on two pieces of tinsel alongside one another for adequate width.

Various other materials are used to form the folded-over type of wing case on nymphs; however, the technique is basically the same, as you'll soon see. This was very much a generic nymph you just tied; you can vary it to your heart's content by changing and mixing and matching any and all components: body, tail, legs ribbing, wing case. Also, you can change its size and shape. Actually, there are few natural nymphs that you can't successfully imitate simply by innovating from this basic design.

But there are some other types of nymph flies well worth learning. Let's tie a pattern that's been a great favorite for over a century on every continent where people fly fish for trout. You'll learn about a most remarkable body material as well as two different wing case materials—one traditional, the other quite new.

There are a couple of other differences of note. The underbody consists of dubbing that matches the color of the pheasant tail. The fly is tied weighted, however, if you wish to take it without the lead wire, that's fine. The ribbing is copper, and you'll have your choice of tinsel or wire, the latter being ideal for smaller sizes.

## Tying Terms

***Reverse-Ribbing*** This means wrapping the ribbing in the direction opposite that of the thread. You'll recall that early in the book, I told you there were a few exceptions to the rule that everything be wrapped in the same direction. This is one of them.

***Adhesive-Treated Feather*** This is a feather that has been impregnated with adhesive, either by spraying or by manual application. I suggest the spray method and the use of fixatives that can be obtained in art stores, craft stores, and some fly shops. By spraying on two light coatings, with drying time between, feathers are rendered tougher and more malleable, and strips from them make excellent wing cases. Goose and turkey feathers are the types most commonly used in this process.

***Bugskin*** This is a relatively new material as of this writing. It is, simply, very finely shaved leather. Here, you'll see how it can be used in place of sprayed feathers for making virtually indestructible wing cases. It comes in a great array of colors, including metallic colors, which allows substitution of this material for the tinsel in Flashback wing cases. For this application, you must obtain the thinnest Bugskin available. It does vary from batch to batch, so you'll need to examine it closely when purchasing. Other uses for Bugskin are too numerous to mention here.

***Pheasant Tail*** This is exactly what it sounds like. I use the term to refer to the large tails from a cock ringneck pheasant. They're used both plain and dyed. The nymph of the ubiquitous western pale morning dun is beautifully imitated with olive-dyed pheasant tail.

### ***The Dressing*** Pheasant Tail (P-T) Nymph

*Hook:* Daiichi #1560; 1XL.

*Hook Size:* For this exercise, size 12 or 14.

*Thread:* Coppery brown or plain brown.

*Tail:* Pheasant tail fibers.

*Ribbing:* Very fine oval copper tinsel or wire.

*Underbody:* Fine lead wire, with coppery brown dubbing over.

*Body:* Pheasant tail fibers.

*Thorax:* Same as underbody.

*Wing Case:* Adhesive-treated feather or Bugskin; dark gray or brown.

*Legs:* Same as tail.

### Tying Steps

1. Tie on up front and wrap to the bend, as usual. If you'd prefer to tie in the ribbing at this time, go right ahead. Or you can do it later, as shown here.
2. Stroke a bunch of pheasant tail fibers—six to eight in all—to a 90-degree angle to the quill, so that the tips are even. Cut them off at the base, gauge the length for a

short tail, and tie them in on top of the hook with a few pinch-wraps. After securing them, gently fold back the pheasant fibers and, if possible, catch them in the materials clip.

3. If you didn't do so previously, tie in the copper-wire ribbing, and form a lead-wire underbody, per the Woolly Bugger tying sequence. Use the thread to make a ramp at the rear.

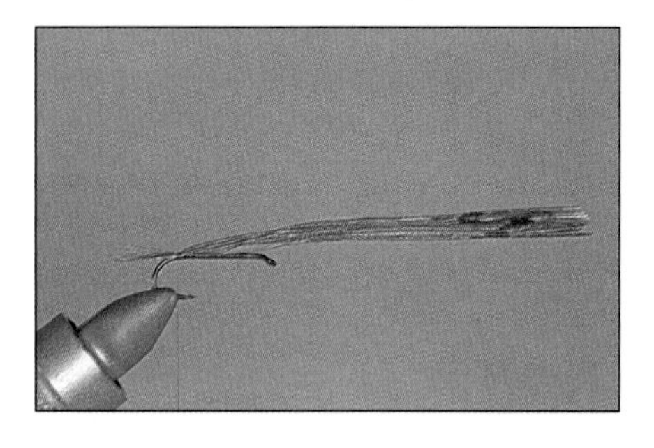

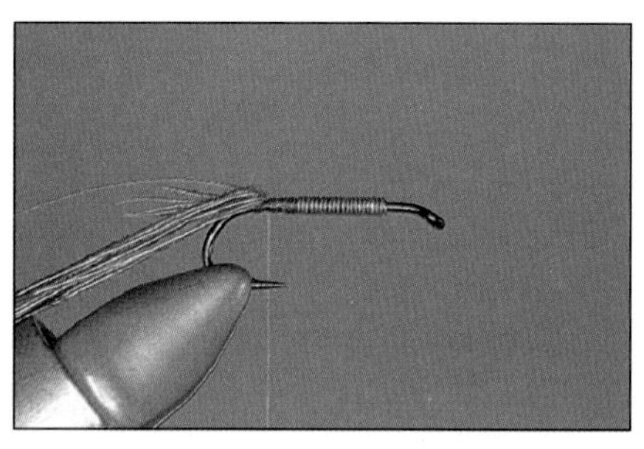

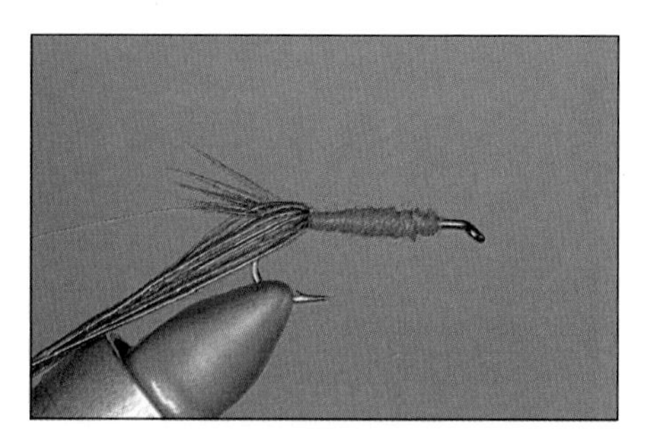

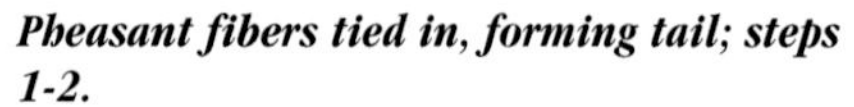

***Pheasant fibers tied in, forming tail; steps 1-2.***

***Ribbing and lead wire in place; step 3.***

***Underbody dubbed into place; step 4.***

4. Using the single-thread method, dub a slim underbody, carrying it up into the midthorax area. Then wrap the thread back into it a few turns, to about 40 percent of the shank length rearward of the eye. This dimension can be varied to create differently proportioned nymphs.
5. Pick up the pheasant tail fibers and wrap them as a bunch, keeping them neat and in line, with no gaps or bumps. You may need to twist them a bit to obtain the desired effect. When you've reached the spot where you left the thread, tie off the fibers and trim the butts.
6. Pick up the ribbing and reverse-wrap it, coming underneath the hook and toward yourself. This is a two-handed operation. Keep the turns evenly spaced, not too close together, and under firm tension. Tie off the ribbing at the front of the body.

***Wrapping the body; step 5.***

7. Here, we'll tie the Bugskin wing case. (The feather version goes on in exactly the same way; it's reinforced with adhesive beforehand.) Cut a narrow strip of Bugskin, as shown. Proportions are important; the strip should be slightly wider than the thorax portion of the body. Taper the front end a bit, so that bulk won't be excessive at the tie-in point. Then tie the strip in as shown, hanging to the rear with the "good" side down.

8. Dub a thorax using the same material that you used for the underbody. Don't crowd the eye.

***Ribbing in place; step 6.***

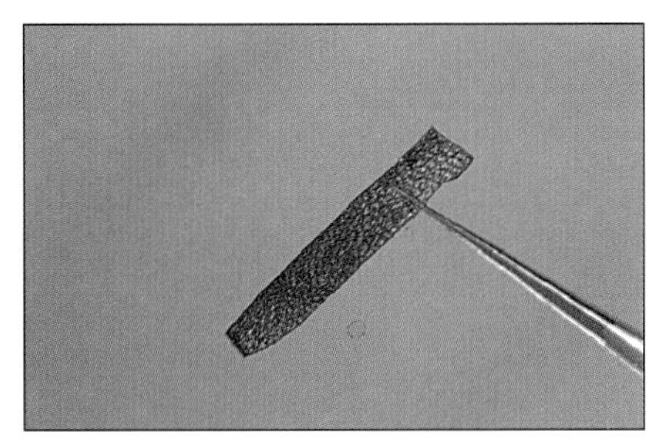

***A piece of Bugskin prepared for the wing case; step 7.***

***Bugskin tied in; step 7.***

9. Stroke a bunch of pheasant tail fibers—about eight in all—to a 90-degree angle from the quill, so that the tips are even. Cut them off, and use the distribution-wrap procedure to form a "hackle" just as you did on the previous nymph.

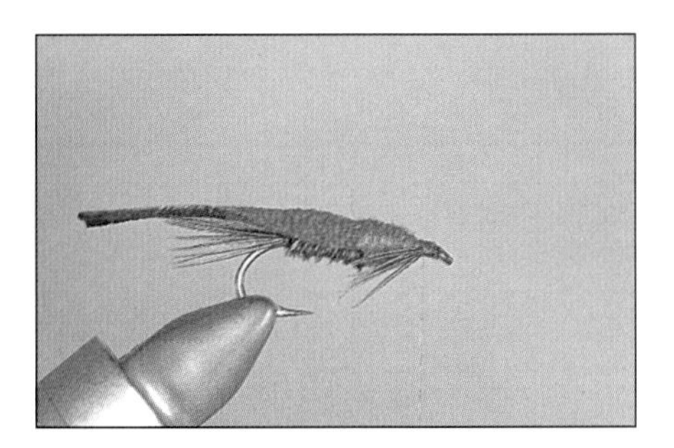

***Thorax and legs in place; steps 8-9.***

***Two views of the completed Pheasant Tail Nymph.***

10. Bring the wing case over, tie it down securely, trim, whip-finish, lacquer the head, and buy a ticket to New Zealand.

The adhesive-treated feather-section wing case, should you opt for it, is executed in exactly the same way as the Bugskin version. Cut out a narrow section of feather; just as with the Bugskin, it should be slightly wider than the thorax portion of the body. Notice that the tip, or outer end, of the section is much finer in texture than the butt; this is the end you'll be tying in. The material will be doubled over later, as was done on the Flashback, so tie it in good-side-down—"good" meaning exterior. Keep the strip centered, and do not allow it to pleat or fold during the tying-in process. Wrap forward and back to secure, and trim off the tag end. Then proceed as described in the pattern exercise.

For the next fly in this section, we'll tie the unlikely Prince Nymph. I've tried to track down the origin of the name, and while I can't come up with a definitive source, the consensus seems to be that it bears the name of the originator, a guy named Prince. I very much doubt that he and the rock star of recent popularity could be one and the same; it has to be some other Prince.

There are two materials in this fly that you'll be encountering for the first time, at least in this book: biots and peacock herl. The latter is one of the All-Time Hall-of-Fame materials. So many great patterns utilize peacock herl that they're beyond counting. I wouldn't want to be without some peacock herl flies in my box, especially not in rainbow trout country.

### *About Peacock Herl*

This material, as you might guess, is obtained from peacocks. The more recognizable source is the magnificent tail that the male of the species likes to fan when showing off. However, the flanks of the bird may be an even more important source, for they yield large quantities of long fronds that, in some respects, are more suitable for fly tying than those from the tail. I should mention that both tail and flank herl are collected during molting; the birds are not killed in order to obtain their plumage.

Usually, peacock tails are sold in small bunches of a half-dozen or dozen feathers. They're commonly graded for quality, which generally relates to the size of the feathers and how luxurious the herl is. That is fine, but there are times when smaller is better, so be aware of what sort of flies you'll be using the peacock for before laying out money for the top grade.

The herl on the tail fronds tends to be the more thick and luxurious. The quills are a little heavier and stiffer, too, which can be a liability or a blessing, depending on whether they are to be wrapped to form herl bodies or tied on in fronds, streamer style. Flank herl is commonly sold in bundles and goes by the name of *strung herl.* The fronds are longer and finer. This is a very economical form in which to buy peacock, provided you're in a position to inspect the bundles. Sometimes the birds from which strung herl is obtained are heavily into molt, and there isn't a lot of herl on the quills. Be sure to check for this.

Even the best peacock herl isn't very durable. In the tying exercise, you'll learn a method for working with herl that effectively reinforces this most valuable body material.

**Tying Term**

***Electronics Clip*** This is the small, spring-loaded device that's used for work on circuit boards. It makes a great herl-twister.

### *About Biots*

*Goose biots.*

This oddly named material is found on the leading edge of the flight quills of larger birds, notably geese and turkeys. Considering their function in life, it's no surprise that biots are quite stiff and sturdy. They're also quite short, which limits their utility somewhat. Biots are mainly used for legs and tails on nymphs and for the wings on the fly we're about to tie. They come both in their natural form and stripped of their herl, which is done with a bleaching process. They're available in an assortment of dyed colors, as well as white.

***The Dressing*** Prince Nymph

*Hook:* Daiichi #1560; 1XL.

*Hook Size:* For this exercise, size 10 or 12.

*Thread:* 8/0 Uni-Thread or comparable; dark brown or black.

*Tails:* Two brown or gray goose biots; gray used here.

*Ribbing:* Fine oval silver tinsel.

*Body:* Peacock herl.

*Hackle:* Brown feather, preferably from a hen cape.

*Wings:* Two white goose biots.

**Note:** The Prince Nymph is dressed in a variety of color schemes. This one happens to be a favorite of mine.

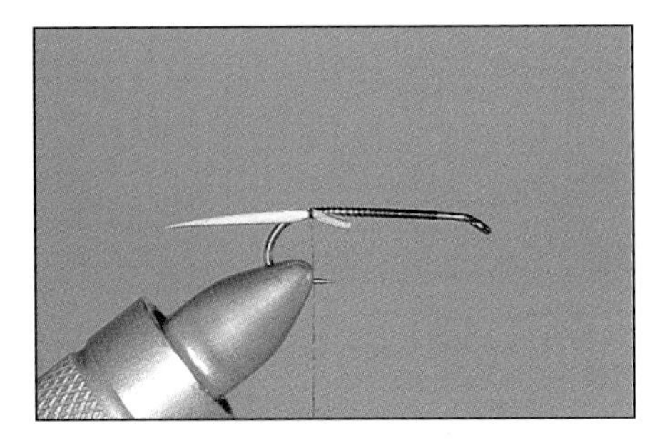

*Forming tails with biots; step 2.*

## Tying Steps

1. Tie on a bit rearward from the eye and wrap to the bend. Notice that I haven't instructed you to tie in the ribbing tinsel, as was done on the Gold-Ribbed Hare's Ear. The reason is that here this would cause interference with the tailing procedure.
2. Tie on the two gray biots. They go on the sides of the hook, front and back, respectively, with the curvature flaring away from the hook. The length should be about equal to the shank. Trim the butts, if required.
3. Now tie in the ribbing tinsel against the far side of the hook, trim its tag end, and secure it in the materials clip.
4. The following is my method for making a peacock herl body. Select four or five fronds whose herl is the proper thickness for the fly you're tying. Trim off a little at the tip ends to eliminate the portion where the quills are so fine that they will break easily when being worked with. Tie in the bunch by the tip ends.
5. Using the technique that was described for making a dubbing loop, run a length of thread from the bobbin that's about twice the length of the herl. Loop the thread around your left forefinger, and bring it back up over the hook. Wrap rearward a few turns, so that both ends of the thread are secure and are adjacent to the herl at the base of the tail.
6. Wrap the thread forward to a point 75 to 80 percent up the shank. Now cut off one side of the loop that you left hanging, and trim it to the length of the herl.

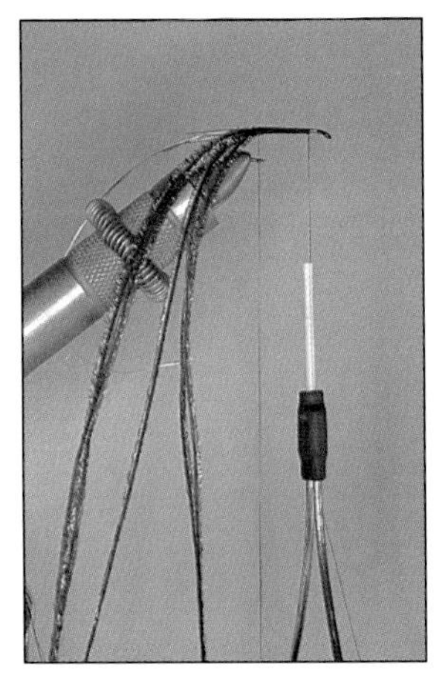

*Ribbing and herl tied in, thread for twisting in place; steps 3-6.*

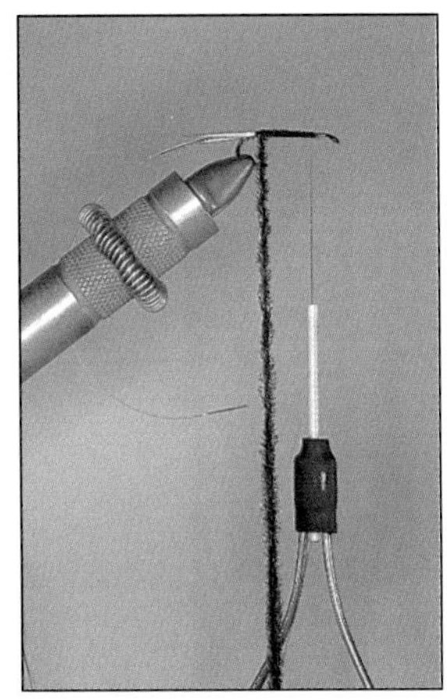

*Twisting the herl together with the thread; step 7.*

7. Twist the herl and thread together, forming a sort of virtual chenille. After a half-dozen twists, take a turn around the hook. Once you've gotten that first wrap in place you can pretty much twist to your heart's content, but don't take too many twists at first; this can cause the herl to break.
8. Keep twisting and wrapping until you reach the spot where the thread is waiting. You may be able to do this all by hand; however, if the bunch of herl resists twisting, don't hesitate to resort to the electronics clip I described earlier. Tie down the herl and trim off the tag ends. Then reverse-wrap the ribbing tinsel.
9. Now you have the option of making either a nymph-type hackle or a "true" hackle, as will be described in the Shushan Postmaster pattern in the streamer-fly chapter. For the sake of simplicity I show

the former here, but I will mention that a wrapped hackle is commonly employed on this fly. Should you wish to give it a try, look ahead to the Shushan Postmaster exercise for instructions. Keep in mind that if you do opt for the wrapped hackle, you'll need to allow slightly more room ahead of the body.

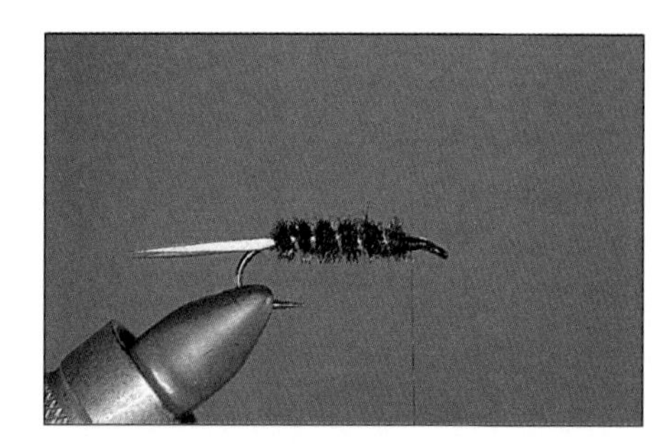

***Wrapping the herl body; steps 7-8.***

***Reverse-wrapped tinsel ribbing in place; step 8.***

10. After making the hackle, trim as required, and take a few housekeeping wraps to create a smooth thread base for the "wings."
11. Now take the two white goose biots and tie them in one at a time, flat atop the hook, tips down, in such a manner that they lie rearward at an angle to the hook shank on each side. They should extend approximately the length of the body, or just a touch beyond. Whip-finish, lacquer, and stash this princely fly in your royal box.

***Beard in place; steps 9-10.***

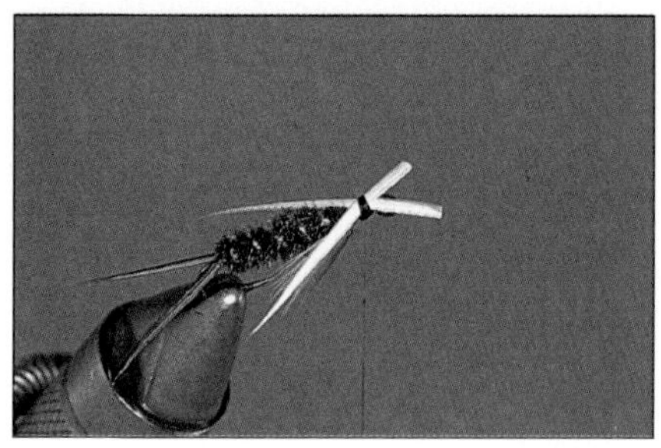

***Tying on and positioning the wing biots; step 11.***

***The completed Prince Nymph.***

I might observe that the folded, wrapped, and stroked-down form of hackle can be used in place of a beard any time you desire. It's especially good for larger flies, such as streamers and salmon flies.

Let's finish this section with an emerger pattern. Emergers are intended to represent just that: emerging forms of aquatic insect life. An awful lot of trout food gets eaten during these fleeting, peril-fraught moments between its nymphal and adult stages, a fact that was not generally appreciated until relatively recent times. Those old Catskill geezers who steadfastly resisted showing me anything of how they fished their wet flies were, mostly without being aware of it, fishing emergers.

There are scores of emerger styles. The one that follows is quite simple, yet it works, and you may never need anything more complicated. The main consideration with emerger flies is how they behave in the water. How you fish them has a great deal to do with this, of course, but so does your selection of materials. Here, you'll learn about one that's made a bit of a splash (no pun intended) in recent years.

*Cul De Canard (CDC)* is the name given to those soft rump feathers at the rear end of a duck or goose. They vary somewhat in texture; some are almost as soft as marabou while others are more like soft hackle—and in fact are used as such. They come in natural shades—white, off-white, gray, and beige—as well as a number of dyed colors.

Much has been written about the floating properties of cul de canard. They were once mainly attributed to the presence of oil, which is applied by the duck during preening. In fact, this was such a marketing hot button that processors of these feathers went to the considerable trouble of reimpregnating those feathers that had been dyed in order to replace the oil that the dye bath had removed.

Ah, but that isn't the essential factor. Turns out the reason the feathers tend to float is that along the barbs are thousands of tiny, microscopic filaments. These are called *barbules.* Their propensity for trapping air is what causes the flotation. It also effectively simulates something that occurs at certain times with natural nymphs and emergent life: the entrapment of air or gas bubblets. This is why CDC flies work as well as they do.

The barbule phenomenon was dramatized in an article in *American Angler* magazine in 1994. The author used photographs taken through a powerful microscope to show the barbules and the entrapment effect. This should come as good news to suppliers, who can now cease and desist with the grease jobs. Actually, the best way to keep CDC floating is to clean the fly of any fish slime by dunking it in a solvent—or simply rinsing it in the stream—and then drying it completely in a desiccant of the type sold in fly shops.

So, why use CDC in an emerger, which is essentially a wet fly (or at least a damp one)? The answer is twofold. First, it looks like an emerging wing popping out of the split wing case of a nymph; second, its buoyancy causes the fly to rise toward the surface and hang in the surface film, as so many real bugs do.

***The Dressing*** Olive CDC Emerger

*Hook:* Daiichi #1560; 1XL.

*Hook Size:* For this exercise, size 10 or 12.

*Thread:* 8/0 Uni-Thread or comparable; brown or olive.

*Tails:* A few wisps of Hungarian partridge.

*Body/Thorax:* Olive dubbing.

*Hackle (Legs):* A few more wisps of Hungarian partridge.

*Emergent Wing:* A clump of grayish CDC.

### Tying Steps

1. Construct the tail and body using the techniques you learned in the preceding lessons. Leave enough space for a prominent thorax.

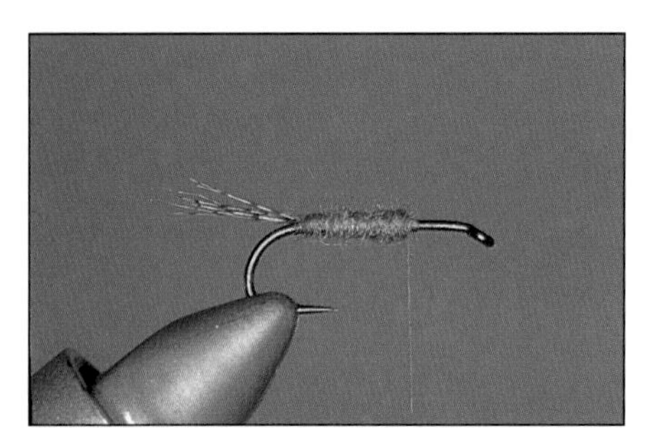

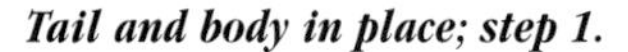

***Tail and body in place; step 1.***

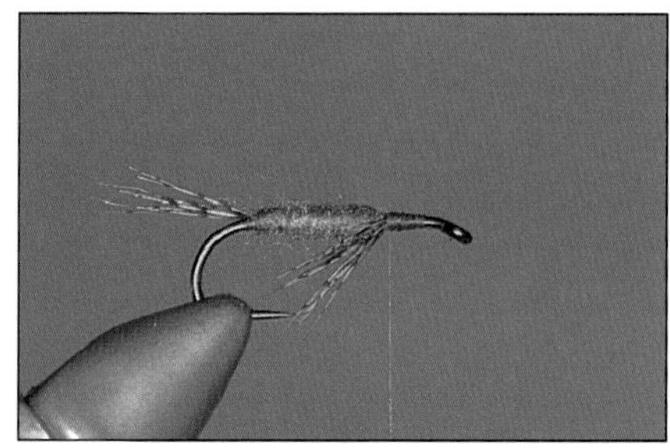

***Legs in place, thread base wrapped; step 2.***

***Wing in place; step 3.***

2. Tie in the legs, and create a little thread base for the CDC.
3. Tie in the CDC in a bunch, as shown.
4. Dub the thorax, whip-finish, and that's it.

***The completed Olive CDC Emerger.***

If you carry a little fine-mesh net on-stream, you'll be able to capture insects in their emerging forms during hatches. This will show you what the bug of the day looks like in that state, and then you can make up your own emerger pattern to simulate it. Lots of fun. Incidentally, there's a product available called the Catch-A-Hatch that consists of an aquarium net mounted on one of those extendable pointers people use in lectures. Very neat.

***A trout after three nymph naturals and an imitation.***

◂F I V E▸

# *Streamers*

An enormous number of diverse flies fall into the general category of streamers. Basically, it encompasses everything tied to represent a forage fish of some sort, as opposed to an insect, crustacean, or the like. The streamer school of fly tying is very much an American tradition and not an import from across the pond. The Brits, from whom we inherited so much of our fly-tying methodology, never seemed to go in for forage-fish imitations.

Many of these flies don't look like little fish; in fact, some of them look more like psychedelic Christmas tree ornaments than anything else. These are the ultimate examples of attractors; of the anything-goes school of fly tying. Amazingly, a lot of these bizarre creations work wonderfully well. Tells you something about the fish mentality.

In this section, you won't do anything terribly weird. What you will do is learn general streamer construction and how to handle streamer-fly materials—some traditional, others contemporary. By mixing and matching items from these groups, you'll be able to create many beautiful and functional effects. To begin, let's learn some terminology and look at the types of hooks on which streamers are tied.

## *About Streamer Hooks*

Some of the following information was set forth in general in chapter 2. Here, I'll get a little more explicit.

Traditional streamers, with certain exceptions, are tied on long-shank hooks. These accommodate the types of materials generally used and contribute to overall shaping. However, in the case of saltwater flies, many of which are streamers of some sort, you'll see regular-shank hooks being used. Oceanic fish don't nip at things; they swallow them whole and keep moving.

Streamer hooks range from 4X to 10X long—those X-words again! The 10X is rather extreme. My preferred hooks range from 4X to 8X in shank length. In fact, I've designed a hook with a 7X-long shank, which I introduce in this book.

Streamer hooks require fairly heavy wire; you'll find that this varies from model to model. The reason, besides helping to sink the fly, is that the long shanks give a large, powerful fish additional leverage, and the wire must be rigid enough to withstand this.

Heavier wire affects not only the weight of a hook, but also the design. On a fine-wire dry-fly hook, a shorter point works just great, because it doesn't take much to cause this little dart to penetrate. The same point on a heavy-wire hook, however, would be too stubby and would have poor penetration characteristics. It would also preclude resharpening the point, as this would shorten it even more. Thus a well-designed streamer hook has a longer, more gradually tapered point.

The barb design is also affected by wire diameter. Too high a barb is another impediment to penetration and can actually work against effective holding. As a powerful fish battles an angler, the barb can enlarge the hole at the point of entry; with a vigorous headshake the fish can send the hook flying. You'll see this in salmon fishing all the time.

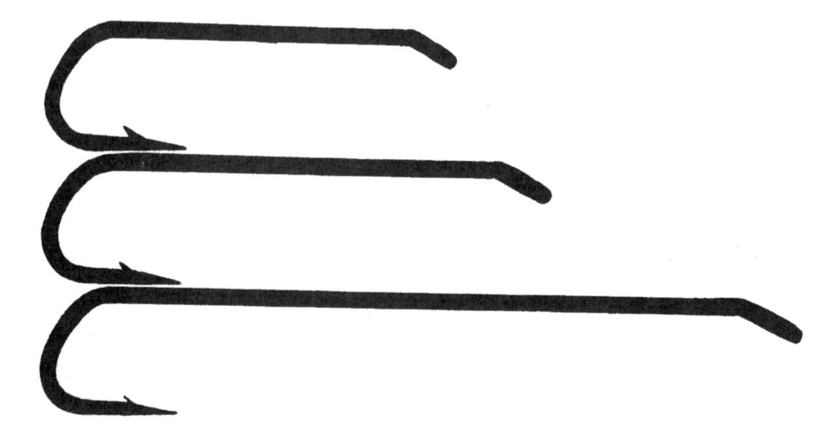

***Some comparative shank lengths of streamer hooks. From the top: 4X long, 6X long, 10X long. Also note the differences in barb sizes.***

The shape of the hook bend is somewhat arbitrary. Actually, any of the common designs will work okay, but a compound curve—as is employed in the variations of the Sproat and Limerick bends—seems to provide a better physical and visual balance. These shapes also accommodate the longer point and throat better than does a semicircular Model Perfect bend.

The front end of the hook—specifically, the eye—also comes in for some scrutiny as to design and workmanship. First, the eye shouldn't be turned down at too severe an angle, because this creates problems for the tyer who's working up near the eye and also affects the positioning of the leader knot. You'll encounter such designs a lot, unfortunately. A deflection of about 30 degrees works well for this type of hook. Straight-eye hooks are also very popular for streamers; I personally happen to favor them.

Another consideration has to do with the design of the eye itself: It can be looped or regular. This is probably a new term for you: *looped eye.* It describes an eye that's formed by running the wire back down the shank a little way, rather than simply shaping it into a ring. The wire that extends down the shank is called the *return wire.* It should taper away to nothing, as illustrated in the drawings.

Either type of eye will work, but I strongly suggest the looped eye for most streamer tying. The slightly wider base provided by the double wire better accommodates the types of materials used for streamers—both feathers and hair. It also eliminates any rough edges or gaps, which are all too often encountered in regular eyes. As a matter of fact, you should carefully examine regular eyes for such defects.

***Examples of looped eyes. The uppermost eye illustrates the undesirable characteristics mentioned in the text.***

There's some additional terminology that applies to streamer flies, and you'll become familiar with it as you work through the tying exercises. One term, however, requires clarification here and now. While streamers don't emulate insect life, the term *wing* is still used; you'll encounter it throughout fly-tying instructional literature. It's a term of convenience. It refers to those materials that are tied on in the position where wings would go on a conventional wet fly. Not descriptive, perhaps, but utilitarian, and I wouldn't want to be the one who tried to change it.

For your first streamer, you'll tie a wonderful traditional pattern: the Shushan Postmaster. Unless you come from a second- or third-generation northeastern fly-fishing household, you've probably never heard of it. The pattern was designed by a gentleman named Lew Oatman, who lived near the New York–Vermont border. One of his favorite fishing buddies was a character named Albie Pringle, who was the postmaster in the nearby village of Shushan—hence the name. The two did most of their fishing on a once renowned river, the Battenkill. It's still there, and so are the fish, but no one can find either beneath the layers of aluminum canoes and rubber inner tubes and the rubbish and litter with which the occupants of same inundate the landscape.

First, a little orientation. This fly belongs in the imitator category, even though it isn't a terribly precise one. Like so many trout rivers, the Battenkill holds several species of minnows that serve as forage fish for the wild strain of predatory brown trout that rules the river. Lew Oatman's pattern was designed to look like a natural baitfish—probably a golden shiner. The components of the fly bear this out: a prominent tail; golden yellow flanks with a hint of glitter; a dash of red at the throat in the gill area; a mottled, brownish back; and prominent eyes. The latter indicate that Mr. Oatman was a keen observer of predatory fish behavior. In recent years, experiences in Alaska and elsewhere have verified that the eyes of a streamer fly prompt strikes.

While I offer this fly as an ideal lesson in streamer tying, I can assure you that it's not simply an academic exercise. This, and quite a number of similar streamer dressings, are time-proven producers over a very wide geographic area—even in those places where the natural forage fish they were designed to imitate don't exist. I've taken fish all over the world on such patterns.

## Tying Terms

***True Hackle*** This refers to a hackle that is formed by folding, wrapping, and stroking down the barbs of a feather. It's similar to the beards you've been tying, but has a fuller, more cornucopia-like deployment, and holds its shape better in currents. It's also an aesthetically pleasing type of hackle and favored by those who dress classic flies for framing.

There are two concerns you must address when using this hackling method. First, the feather must be the proper size (meaning barb length) for the fly you're tying. Second, the quill must be very fine, so that only minimal bulk results from the wrapping. This is critical when you're subsequently going to tie on a wing.

***Squirrel Tail*** This is often used in streamer tying. As pure protein, it's one of the more unforgiving materials: hard, slippery, and resistant to being compressed. Thus, it needs some special handling, which essentially means using discrete techniques for tying it in place.

When you're working with squirrel tail in any appreciable quantities, I strongly suggest you tie it on in small bunches, in two or three stages. This gets lots of thread interlayered. You may also wish to put a droplet of an effective adhesive on each set of thread wraps as you go. Dave's Flexament is a good one, as is Shoe Goo, thinned with Toluol.

Several varieties of squirrel tail hair are used in streamer-fly tying and in the closely related school of hair-wing salmon-fly tying. Among these are the hair of the common gray squirrel; the black, or melanistic squirrel; the red squirrel and its slightly larger close relative, the pine squirrel; and the large and handsome fox squirrel that's so populous throughout Ohio, Indiana, and that area. The fox squirrel composes the wing in the pattern we're about to tie. It's the hardest of all the hair varieties to handle, so if you can deal with it, you'll have little trouble with the others.

***Jungle Cock*** This refers to the highly individual feathers from the neck of an Asiatic bird, the proper name of which is the India gray jungle fowl. I might also mention the Latin taxonomy, *Gallus sonerati*—not for the sake of getting technical, but simply to accurately establish the bird's identity.

There's been a lot of controversy over old *Gallus.* In 1967, the import of these very popular feathers was banned, for reasons that are still not clear, as the bird is not threatened. In fact, it's quite populous throughout large areas of Southeast Asia and hunted by the natives for food. This means that the pelts are actually a by-product. Still, their import is closely controlled.

In recent years, the Department of the Interior has granted import licenses to several overseas suppliers who raise these birds in a semidomestic state. Thus, a limited and quite pricey supply of this commodity is once again legally available. Most of the legal jungle cock is sold in packets of ten matched "eyes." As of this writing, they run $.60 per eye, so a pair of eyes for a fly like the Shushan Postmaster costs $1.20! Don't panic; I'll show you a couple of viable alternatives.

On the hair-wing fly we're about to tie, we'll resort to a painted eye. This is well suited to hair-wing streamers, on which the wing material inevitably results in a built-up head—prerequisite to a painted eye.

***The Dressing*** Shushan Postman

*Hook:* Daiichi #2340 or #2370 streamer or comparable.

*Hook Size:* For this exercise, size 4 or 6.

*Thread:* 8/0 Uni-Thread; first yellow or something pale, then black.

*Tail:* Mottled turkey, tied on like a wet-fly wing.

*Ribbing:* Narrow flat gold tinsel.

*Body:* Golden yellow floss.

*Hackle:* Soft red hen or rooster hackle.

*Wing:* Fox-squirrel tail.

*Cheeks (Eyes):* Jungle cock or substitute; here, a painted eye.

### Tying Steps

1. Tie on a little way to the rear of the eye, and wrap to the bend. If you're using a hook with a looped eye, tie on over the rear end of the double wire and wrap rearward until you've cleared it by a few turns. Even the best-made looped-eye hooks may have a little gap between the main and the return wires; this closes it.
2. The tail on this fly is tied in exactly the same manner as the wing on the wet fly from chapter 3. Cut out two matched sections from opposing mottled turkey quills and cup them together, concave-to-concave. Be conservative in width. Tie them in place at the bend with a few pinch-wraps followed by several securing wraps.

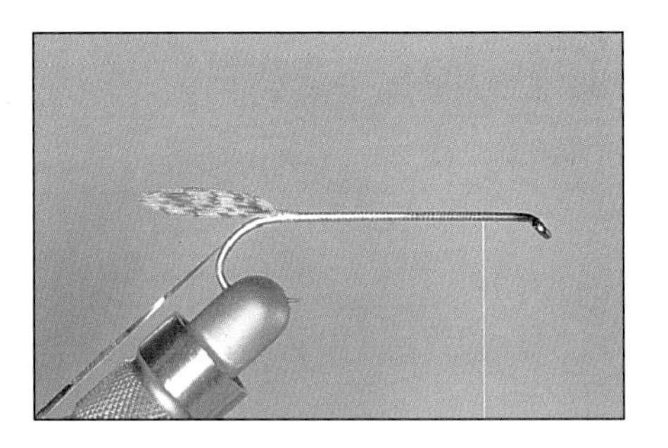

***Tail in place, ribbing tied in, thread advanced forward; steps 1-4.***

3. Tie in the ribbing tinsel, gold-side-out, at this point. Leave just enough space between it and the tail for one or two wraps of floss; this helps set up the tinsel for the first turn.
4. Trim the tail butts on a slope, then bind them down neatly and tightly, working orward up the hook. Smooth out and equalize the underbody as much as you can.
5. Near the front of the hook, tie in a piece of fairly stout floss. Check the photo for the proper tie-in point. If you're using a looped-eye hook, tie in the floss just off the end of the return wire. If your floss is the four-stranded type, tie in all four. If you happen to have only fine floss, use multiple strands—as many as are appropriate to obtain the thickness you desire.

**Note:** Floppy materials like floss, tinsel, and yarn can be tricky to tie in. Here's a helpful technique: Hold the material between your two thumbs and forefingers and bring it up underneath the hook, just in front of the thread. Secure it against the backside of the hook with the middle finger of your left hand. Bring the thread up and over, and you've got it.

6. After securing the floss, trim the tag end. Wrap the floss neatly to the bend, overlapping the wraps a bit. During wrapping, the floss should flatten, but it should not break down into individual hairlike strands. If you see this happening, twist the floss a little to bring it back into conformity.
7. At the rear, bring the floss behind the tinsel one or, at most, two turns. Then wrap forward to the tie-in point and a turn or two beyond, working the thread forward, too, if necessary. This builds in a little forward taper. Tie off the floss, and trim the excess.
8. Pick up the tinsel and create the ribbing. Keep the turns even and well spaced. At the front, tie off on the bottom of the hook, secure well, and trim off the tag end.

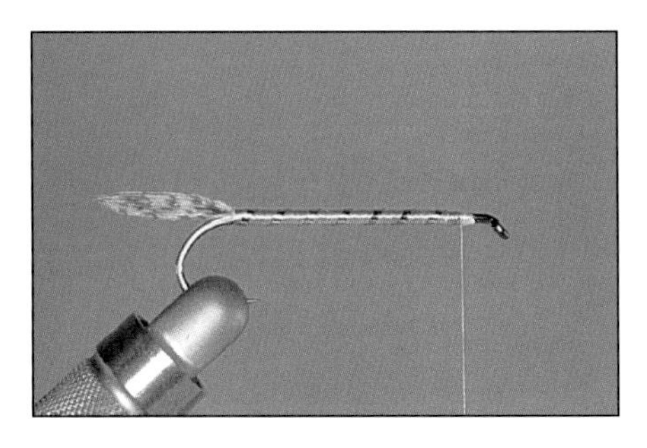

***Body and tinsel in place. This is similar to the Professor, but larger; steps 5-8.***

9. Now you have the option of making a throat or simple beard hackle, or a wrapped-and-stroked-down hackle, which in traditional British parlance is referred to as a "true" hackle. This being your first streamer exercise, let's stay with the beard for the moment. You'll learn the other technique, along with the marabou-wing option, subsequently. Form the beard out of a soft, red-dyed feather, as shown. At this time switch to black thread.

10. From a fox-squirrel tail, cut a fairly slender bunch of hair, keeping in mind that the wing will be built in two layers. If there are any wild-looking hairs sticking out, dispose of them, but don't stack the hair; streamer wings should taper when wet and not resemble paintbrushes.
11. With your left hand, hold the bunch over the hook, and gauge length. The hair should extend just beyond the end of the tail and no farther.

***Beard in place, thread changed to black; step 9.***

***Gauging wing length; steps 10-11.***

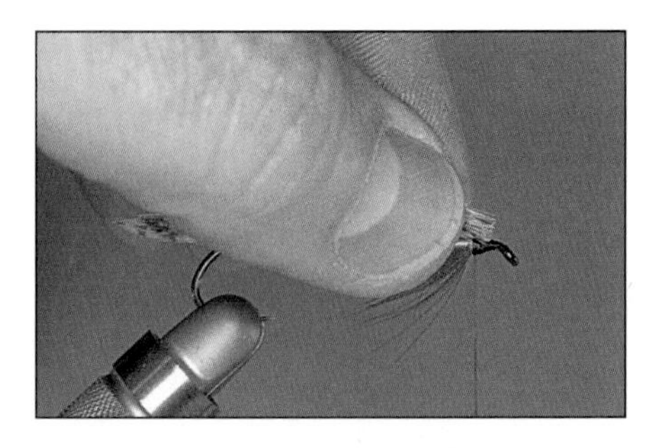

***Hair bunch precut and ready for tie-on; step 12.***

12. Now you have two choices: You can simply tie on the hair and trim later, or you can precut. I advocate precutting. Here's the procedure: Hold the winging bunch exactly in position; trim off the butts, neatly and on a slight taper, just behind the eye.
13. Set the hair on top of the hook and tie it on with plenty of firm wraps, working back and forth. While securing the first layer, you're also creating a thread base for the second.

***First bunch of hair tied in place; step 13.***

***Second bunch of hair being tied on; step 14.***

***Head built up, ready for eye; step 14.***

14. Repeat the process, precutting and all. Secure the second layer, then build a smooth, good-sized head with the thread. Tie off.
15. For the painted eye, you'll need yellow lacquer, black lacquer, and two round applicators, one considerably larger in diameter than the other. The best I've found are the back ends of a couple of drill bits. A round toothpick that's cut in half, then cut again halfway down the tapered part, also works nicely.
16. Apply two coats of clear lacquer, allowing each to dry thoroughly.
17. Have some yellow lacquer handy and the larger of the two applicators. Shake the lacquer bottle well. Lay the fly on its side with its head hanging off the edge of some-

*The yellow dot painted on; steps 15-18.*

thing—a book, perhaps. This position will allow you to paint the far side of the head once you've applied a droplet of wet paint to the near side.

**18.** Dip the applicator into the lacquer, catching a generous drop. The way this works is that as it's brought down onto the side of the head, the applicator causes the lacquer to spread out in a perfect circle. Actually, the applicator doesn't quite touch the head itself, or if it does, just barely. Paint both sides, and set the fly aside for complete drying.

**19.** When the yellow lacquer is dry, use the smaller applicator to make black pupils in the centers of the yellow dots. Allow everything to dry thoroughly, then coat with clear head cement, or use the epoxy process described in the Dick's Dace exercise later on in this chapter.

*Black pupil in place, completing the Shushan Postmaster; step 19.*

*The original dressing, with jungle cock eye. Note that it's tied short, and flat against the thread that forms the head. Refer to the feather-wing sequence that follows for details.*

## The Hackle-Wing Option

The true classics of the Golden Age of Streamers were tied using hackle feathers for wings. While some contemporary streamers, such as the Clouser Minnow, the Woolly Bugger, and others, may be generally more effective, these old flies still work very well, and when skillfully tied, they are absolutely gorgeous. Even if you don't live in an area where streameres are part of the local fly-fishing cult, I feel it's of value for you to know how feather streamer wings are tied.

The main challenge today is finding feathers suitable for the wings on these flies. Chicken growing has changed, and so have chicken feathers. We don't see much of the large, densely webbed stuff of years ago. Mostly, today's streamer tyer must rely on what trickles in from overseas.

Cape feathers are more likely to have the desired properties for streamers than are saddle for the following reasons:

- They're wider and webbier and hold their form better when wet.
- The quills are usually more amenable to tying. Saddle hackles tend to have very fine quills and, under thread pressure, they're liable to roll into any position except the one you want.
- Saddle hackles often have reverse curvature, which results in a messy-looking wing.

Capes, or necks, for streamer tying can be purchased as-is or the feathers can be bought in bundles strung together by thread. Saddle feathers intended for streamer use are almost invariably sold in bundles. In either case, you get a lot of feathers for not much money; the question is how suitable they are. Sometimes, bundled feathers are twisted and not usable at all. Often, a moderate number of nice feathers is bundled with a larger number of ratty ones. Even with good bundles, count on a certain amount of waste.

When you're tying feather wings, consider two factors: First, feathers seat better on looped-eye hooks. Second, you don't need to allow as much space up front; feathers can be securely tied on without covering a lot of head area.

I should also point out that the wrapped hackle you'll see in the marabou-wing version can cause problems if you're new to feather wings, because passing the wraps of quill over the hook may cock the wing feathers. With experience, you'll learn to compensate for this. For now, I suggest that you stay with the simple beard method for making the throat.

So, with apologies to Lew and his pattern, I offer the following simple alteration. The feathers I've selected, brown with a black center stripe, are known as *furnace.* I've chosen them because they simulate the brown/black shading of the squirrel hair. They're quite common, but if you don't want to go chasing around after them, just use brown. Here's the process:

1. Extend the body a little farther forward, as I mentioned.

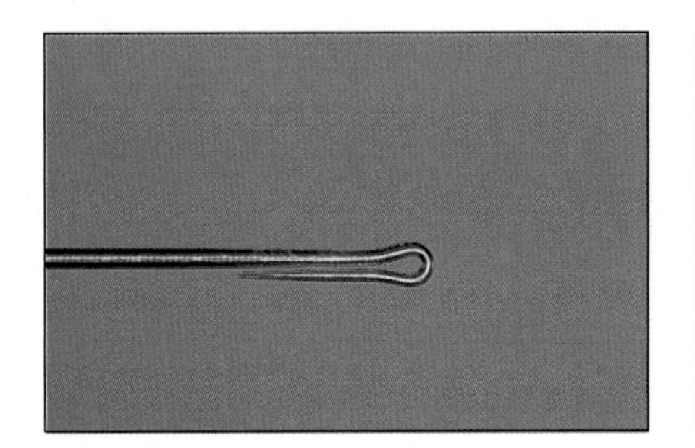

***A looped-eye hook, top view.***

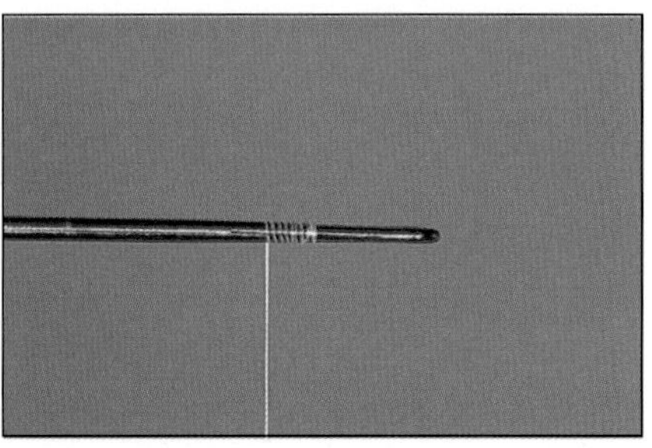

***Start the thread over the double wire, to bring them together.***

***Tail, body, ribbing, and beard done. The fly is ready for the wing; step 1.***

2. Select two closely matched hackles from the rear of a cape. They can both be straight, or one can curve one way and one the other. The only no-no is that they shouldn't both curve the same way, as you'll place them back-to back, and they must complement each other.
3. Place the two feathers dull-side-to-dull-side, with the tips perfectly aligned. Working on both at once, strip off the fluffy barbs near the butts until the feathers approximate the length of the hair wing you just tied. Then remove a few extra barbs from the bottom of the quill, right at the tie-in point. This helps the wing fit up against the front of the body and lie flat, or reasonably so.

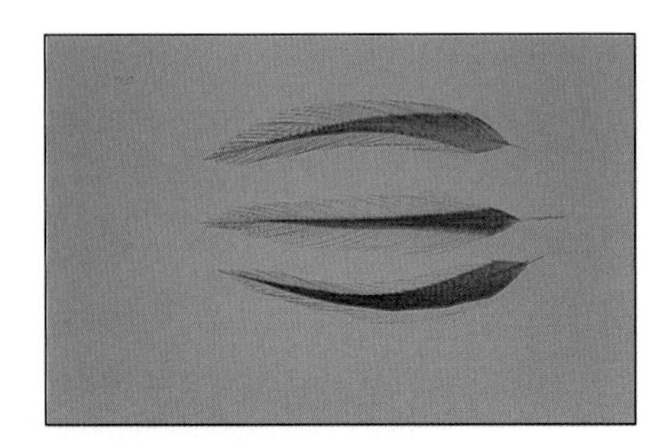

***Typical streamer-wing feathers.***

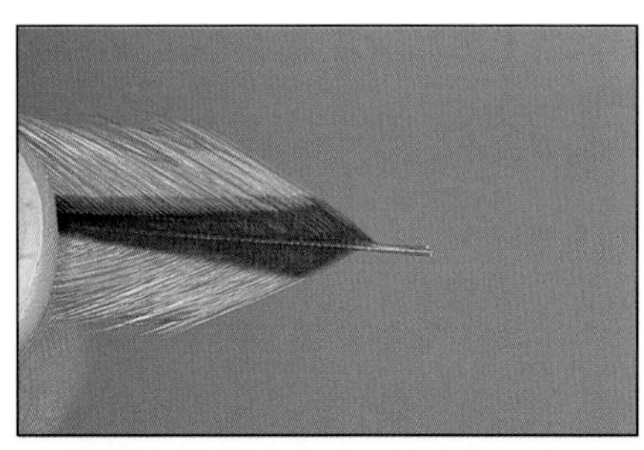

***Wing feathers prepared for tie-in; steps 2-3.***

***Wings in place; step 4.***

4. Set the feathers on top of the thread base, center the quills, and tie them on with a series of very firm wraps. Trim the quill butts, and cover the trimmings with some housekeeping wraps.
5. If you have real jungle cock, here's what to do: Strip off the feathery material on each side of the eye. Then hold the eye flat against the side of the head, and pass the thread over the little delta of feather at the front. Don't tie on jungle cock by the stripped quill alone; this weakens it and may cause it to skew during tying.

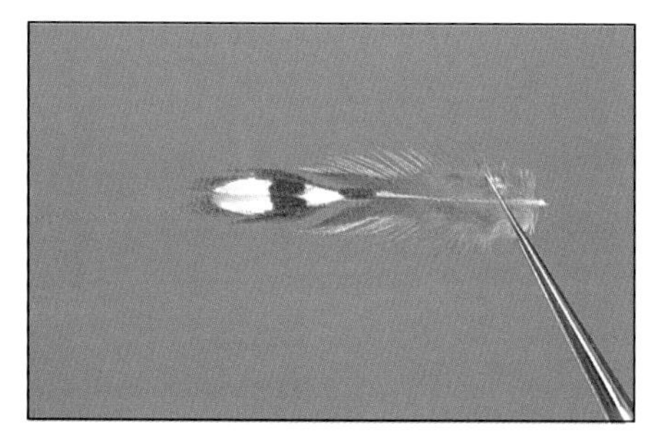

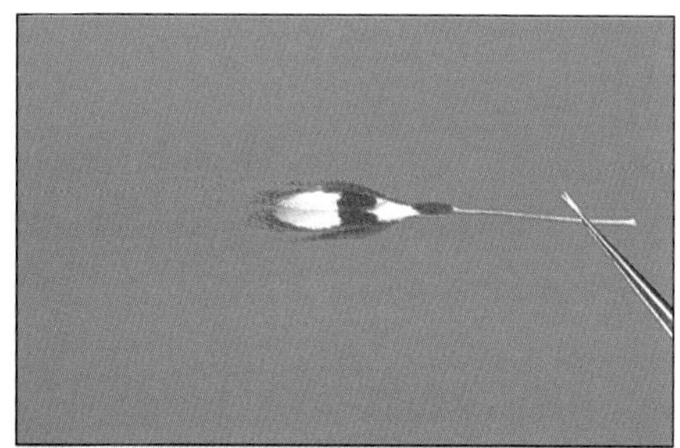

*A jungle cock "eye," before and after being manicured; step 5.*

*The completed Feather-Wing Shushan Postmaster.*

6. Make a mirror-image jungle cock eye on the far side of the head. Secure them both, and trim off the quills. Neaten and smooth the head with some housekeeping wraps, as needed. Whip-finish, and apply at least two coats of lacquer, drying thoroughly between each coat. You now have a very beautiful, if somewhat costly, streamer fly.

I might mention that many of the great traditional feather-wing streamer patterns, especially those in larger sizes, specify a wing made of four feathers, rather than two. The procedure is the same, except that you must take care pairing up the fronts and backs.

## The Marabou-Wing Option and the "True" Hackle

You learned about marabou in the Woolly Bugger exercise. Marabou is a popular and effective material for winging streamers and is frequently substituted for the original hair wing. Let's do that here and, at the same time, learn two more techniques: forming a beard by folding, wrapping, and stroking down a hackle; and making a jungle cock substitute out of common guinea fowl.

1. Tie the Shushan Postmaster just as before, until you're ready for the beard.
2. Pick out a soft, webby, red-dyed hen or rooster hackle with barbs about the length of the tail and as fine a quill as possible. Strip the waste material off the butt end. Prefold the hackle a little, using the same stroke as in the barb-gathering procedure.
3. Tie in the feather at the front of the body by its tip end, as shown. Begin to wrap it and, as you go, use your left thumb and forefinger to stroke back the barbs with each turn. The idea is that the quill becomes the leading edge. You'll only get a few turns, but that's all you'll need.
4. Tie off the hackle feather, trim off the butt, and take a few housekeeping wraps. Now stroke the barbs downward, letting them be swept down to whichever side of the

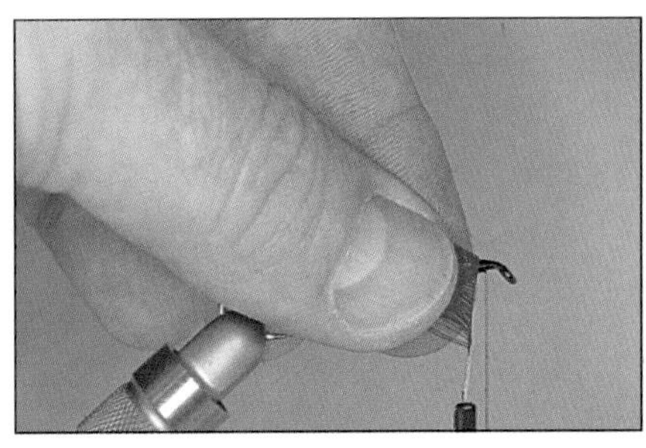

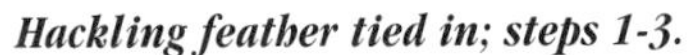

***Hackling feather tied in; steps 1-3.***

***Wrapping the hackling feather; step 3.***

hook they naturally want to go. The few in the center that don't want to go to either side are pruned out with a tweezer. Resist the temptation to cut them off; that would leave little stubs sticking up. If the hackle flows rearward and downward as that in the photo does, you're done with it. If it doesn't, stroke it firmly in those directions, applying pressure at the vortex. Tweeze out any barbs that simply won't behave. If the fibers still resist forming a neat beard, hold them in the position you desire and apply a tiny droplet of Zap-A-Gap on top, where the feather is wrapped around the hook. Rub with a toothpick until dry.

5. Select a medium brown marabou plume with fairly long, soft fibers. Strip off the very fluffy material nearer the butt end. Inspect the tip portion and, if the center quill runs well out toward the end, go in with the points of your scissors and snip it out.

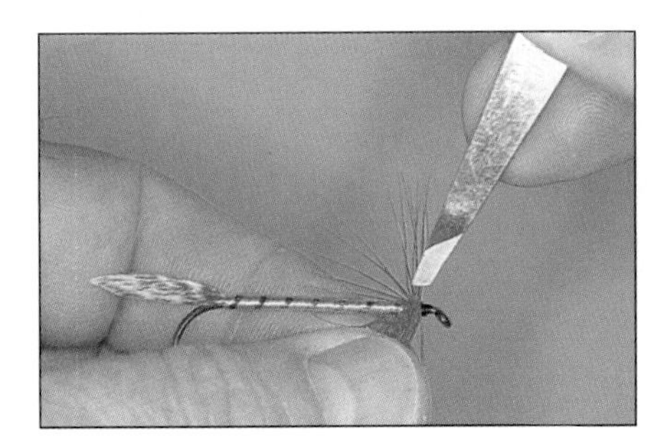

***Tweezing out unwanted barbs; step 4.***

***Hackle stroked down to form "beard"; step 4.***

***Marabou wing in place; steps 5-6.***

6. Tie on the marabou in essentially the same manner you used for the Woolly Bugger tail. Stroke the plume into a neat shape, dampening it a little, if necessary, to make it behave. Tie it on top of the thread base and secure it with a series of firm wraps. The length should be about the same as for the squirrel tail hair. Stay far enough back from the eye to allow for trimming; marabou doesn't trim neatly, as squirrel does. Trim carefully and thoroughly, and cover the trimmed butts with some housekeeping wraps.

Probably, you won't want to go right out and buy jungle cock. Here's how to make an acceptable substitute from an inexpensive feather, the common guinea fowl. These birds have two types of body feathers: single-dotted and double-dotted. The double-dotted type is often used in salmon-fly dressings. The single-dotted is the type we'll use here. The procedure is as follows:

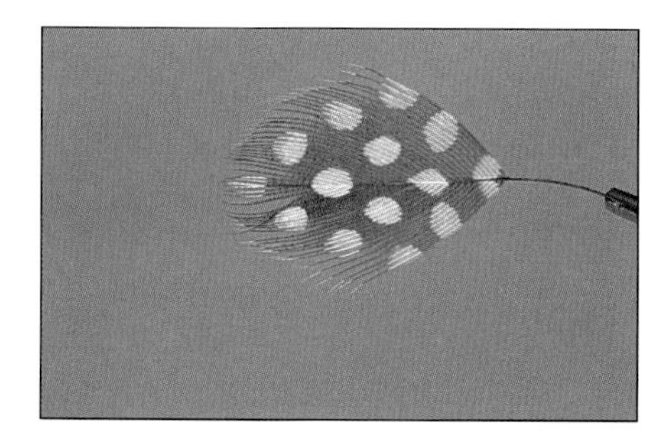

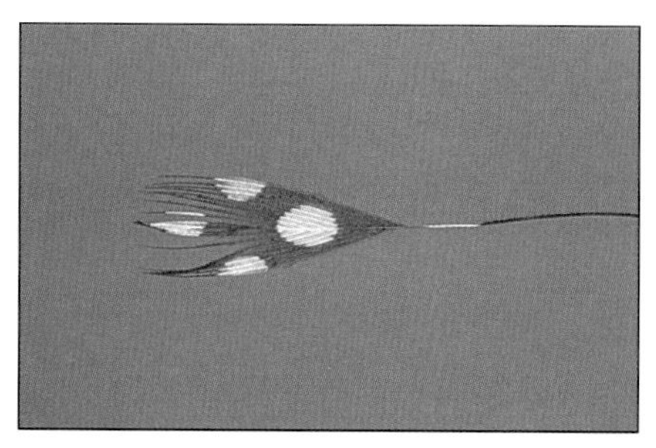

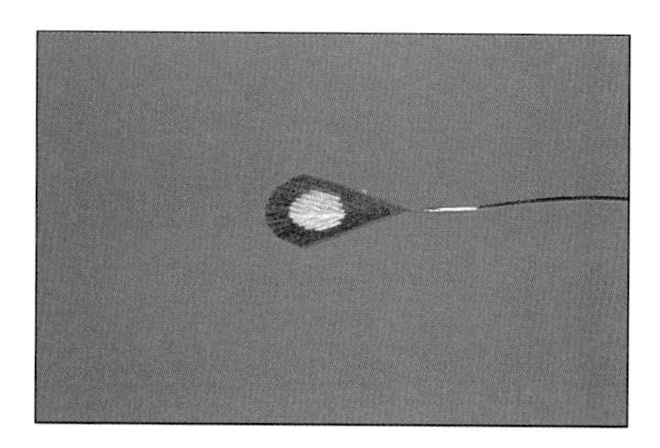

***Making an imitation jungle cock eye out of a guinea fowl feather.***

***The completed Marabou-Wing Shushan Postmaster.***

1. Select a feather with a white dot centered on the quill, near the tip.
2. Strip the black barbs off each side until only one or two remain on each side of the white dot.
3. Trim the end of the feather into a shape that generally resembles a jungle cock eye. Curved-blade scissors are helpful here.
4. Tie the imitation in place as you would real jungle cock.

## The Black-Nosed Dace and Dick's Dace

Now I'd like to take you through another excellent traditional streamer pattern, the Black-Nosed Dace (BND). It was designed by well-known Catskill author Art Flick to imitate a minnow common to the greater Northeast and Midwest. Interestingly, it's also a very effective pattern on the Big Horn, a western river that has no natural black-nosed dace. Does that make it an attractor pattern in that region? A rhetorical question.

Before we start, a little orientation. As I mentioned, this fly is an imitator: It was designed to look like a specific baitfish. The natural black-nosed dace has a whitish belly with tiny, sparkly scales, a tan or brown back, and a conspicuous black stripe down each lateral line that extends from the fork of its tail to the tip of its nose—hence the name. The idea is to use the fly materials in such a manner that they look like the Real Thing. The most critical factor in achieving this is preventing the three colors of hair from mixing; you must keep them in discrete, layered bunches. I'll show you how.

This fly has been shown in many fly-tying books over the years, including one of my own. I repeat it here because it provides such a great exercise in working with streamer hairs. Afterward, I'll teach you a contemporary version that I call Dick's Dace.

### *About Bucktail*

The term *bucktail* refers to a streamer that is, or was originally, tied with—well, bucktail! The one we're about to tie and the Mickey Finn are the two most famous examples. Today, the term still applies to this type of construction, even though many different kinds of hair are used in the tying, including synthetics. *Bucktail* also refers to deer tail, per se, in tying-material parlance.

While the bucktails we use all come from the white-tailed species of deer, the tails themselves vary considerably in a number of respects. This is mainly due to regional factors such as climate and diet. Genetics also plays a role, as do age and sex; some bucktails are actually doe tails.

Large tails with coarse hair are fine for saltwater tying and that sort of stuff. For smaller flies, however, such as this one, the finer and straighter the hair, the better. Remember this when you're selecting tails. The inveterate streamer tyer grades tails as carefully as the dry-fly addict grades hackle.

Bucktails vary greatly in the amount and quality of brown hair they contain. If you intend to tie lots of BNDs, look for tails that have plenty of long, straight brown hair. If this turns out to be a problem, feel free to substitute. Certain bear hairs are the right color for the BND, as are certain squirrel tails.

The term *tag* refers to any small embellishment at the rear of the fly. Here, a little tag of reddish yarn is used to represent the anal vent of the natural minnow.

***The Original Dressing*** Black-Nosed Dace Bucktail

*Hook:* Long-shank streamer; refer to Shushan Postmaster.

*Thread:* Black.

*Tag:* Fine red yarn.

*Body:* Flat silver tinsel.

*Wing:* Three layers; from the lowermost: white polar bear hair, black bear or skunk tail hair, brown bucktail.

> **Note:** I suggest you make a few substitutions. It would be risky to try to tie an authentic BND at this time, as polar bear hair is illegal. We'll use a synthetic substitute. I also suggest substituting embossed tinsel for the flat, and black-dyed squirrel tail for the bear or skunk hair.

### Tying Steps

1. Tie on near the front, and wrap neatly about 25 percent of the way down the hook shank.

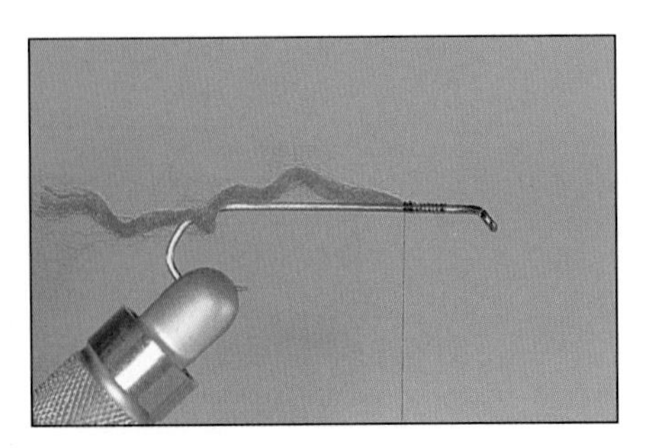

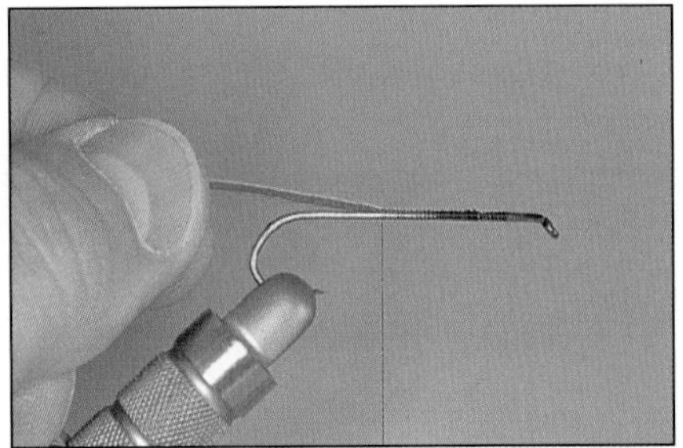

***Tying on the yarn for the tag; steps 1-3.***

2. Tie a short piece of bright red or hot orange yarn on top of the hook, in the position shown.
3. While pulling on the yarn from the rear to keep it tight, cover it with thread wraps, working rearward to the bend.
4. Cut the yarn to length, leaving just a small tag end. Then wrap neatly forward, making a smooth underbody. Stop well short of the eye.

5. Tie in a piece of tinsel, secure it well, and trim off the excess.
6. Wrap the tinsel to the rear and back. The trick to obtaining a smooth tinsel body is to first slightly overlap the wraps; then slide them off the edge, so they lie contiguous to one another. When you reach the tie-in point, remove all but one of the tie-in wraps, take one last turn of tinsel, tie it off, and trim the tag end.

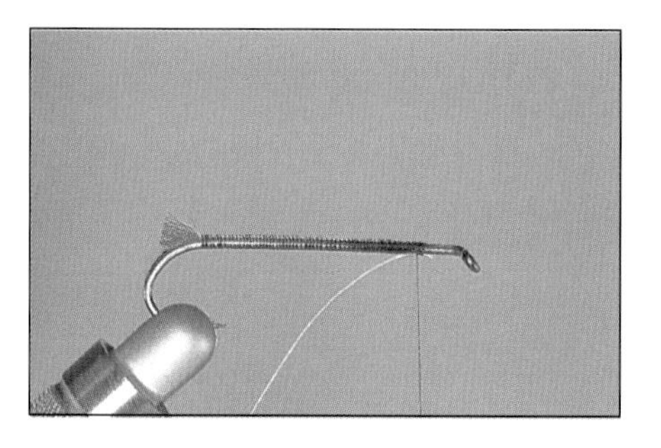

*Tag cut to length, tinsel tied in; steps 4-5.*

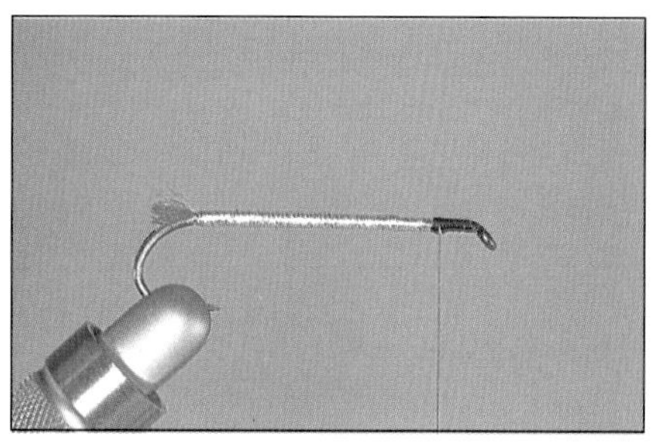

*Tinsel body completed; step 6.*

7. Run the thread to the eye and back a couple of times, creating a thread base for the hair to follow.
8. Cut off a small bunch of the white hair and clean out all shorties and underfur. Tie the hair directly on top of the thread base—precutting or not, as you prefer. Bury the butt ends beneath thread, creating a new base. This is how the layers of hair are kept from mixing. Refer to the illustration for proper length.
9. Prepare a small bunch of the black hair, about the same length as the white or a tad shorter, and tie it on top of the thread base. This is the same procedure used on the Shushan Postmaster, except that here the hairs are of two different colors. Once again, bury the butt ends beneath thread, creating yet another base.

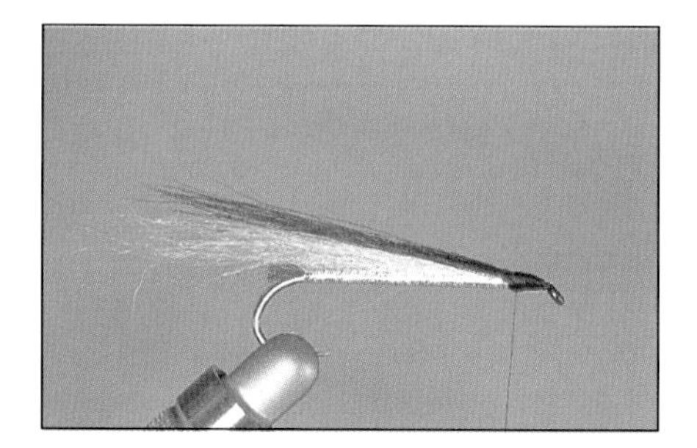

*White and black hairs tied on; steps 7-9.*

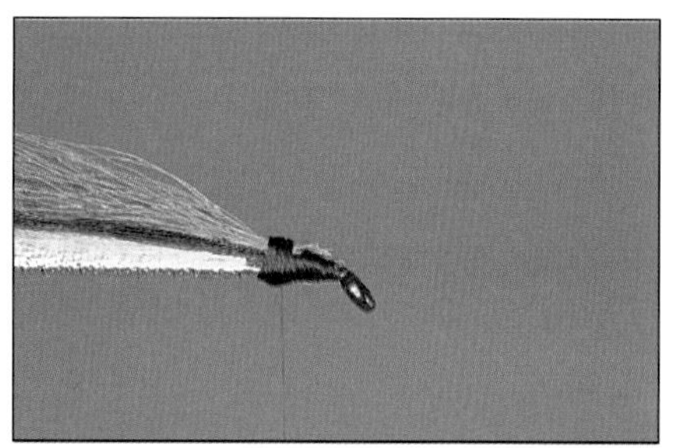

*Brown layer tied on. Note the layered effect; step 10.*

*The completed Black-Nosed Dace Bucktail.*

10. Prepare a bunch of brown hair approximately equal in quantity to the white and black combined. Tie it atop the new thread base, manicure the head with housekeeping wraps, whip-finish, lacquer, and go looking for a large, carnivorous trout.

As much reverence as I have for tradition, and for Art Flick in particular, I'm going to alter the original dressing in several significant respects. Modern materials enable me to do this, and if Art were still around I'm sure he'd be doing the same things.

## Tying Terms

***Tubing*** This refers to a type of material that, like many items used in fly tying, is a knockoff from the decorative-sewing business. Tubing comes in a wide range of sizes and colors. As the photographs will show, it consists of a woven Mylar or plastic tube with a fiber core. For this fly, you'll be using narrow pearlescent tubing.

***Stretch Nylon*** This is a modern, flosslike material. It has certain advantages over conventional floss: It's tough, it's colorfast when wet, and it wraps easily and smoothly. There are at least two brands on the market and quite an array of colors, including some fluorescents. It can be used as a floss or yarn substitute and also in unique ways, as we'll do here. We'll be using it in a more conventional manner later on in the book. I absolutely love this stuff!

***Ghost Fiber*** This is the brand name of a soft, synthetic hair. It works nicely for what you're about to do, but please don't get hung up on the name. There are many similar products on the market these days, including some that will certainly work just about as well. For all I know, there's something even better out there and I just haven't seen it yet. So feel free to substitute.

***The Dressing*** Dick's Dace

*Hook:* Daiichi Model #2340 or #2370 streamer or comparable.

*Thread:* Hot reddish orange; also black Uni Nylon Stretch.

*Underbody:* White Uni Nylon Stretch.

*Overbody:* Narrow pearlescent tubing.

*Tag:* The reddish orange thread.

*Wing, First Layer:* White Ghost Fiber or similar soft hair.

*Wing, Second Layer:* Black-dyed squirrel tail.

*Wing, Third Layer:* Brown bucktail or light brown bear hair.

*Eyes:* Mylar sheet with epoxy over.

*Necessary Ingredients:* Clear epoxy; Zap-A-Gap superglue or the equivalent.

**Note:** Three bobbins are necessary for this dressing, unless you want to wrap the white stretch nylon by hand—which is perfectly okay.

## Tying Steps

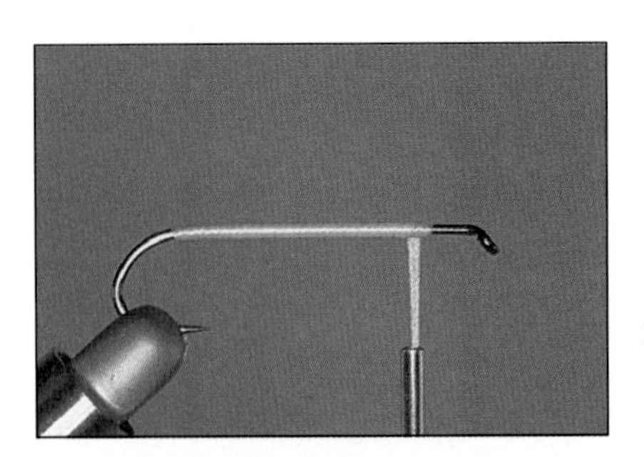

***Body formed of stretch nylon; step 1.***

1. Mount the white stretch nylon in a bobbin, tie it on as you would thread, and apply several layers, working to the bend and back. If you notice that the material is getting twisted and going on "ridgey," spin the bobbin counterclockwise to untwist, as required. When the underbody is complete, tie off with a three-turn whip-finish. Stay well clear of the head area, as shown.
2. Tie on at the rear with the hot reddish orange thread. Remove the fiber core from a piece of pearlescent tubing that's long enough to cover the floss, with enough left over to secure. Run the tubing over the eye of the hook, then over the underbody, leaving a little tag hanging out at the rear.
3. Tie down the rear end of the tubing with the thread; this also forms the reddish tag. Trim off the excess tubing. Secure with either a long-loop whip-finish or a coat of Zap-A-Gap over everything, and rub dry with a toothpick.
4. At the front, tie on with black stretch nylon mounted in a bobbin. Secure the front end of the tubing and trim off the excess.

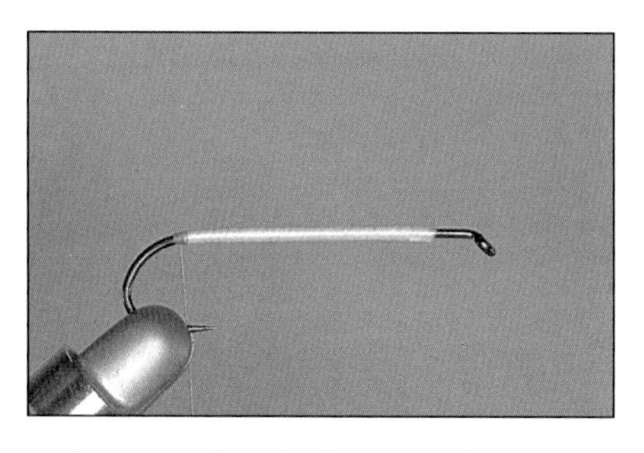

***Body material tied off, orange thread tied on; step 2.***

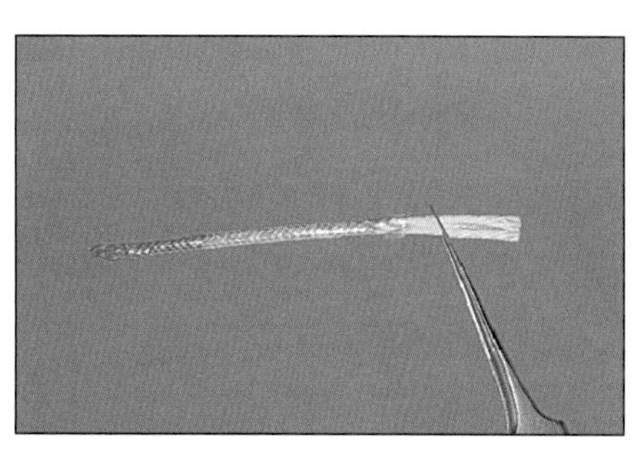

***Removing the core from a section of tubing; step 2.***

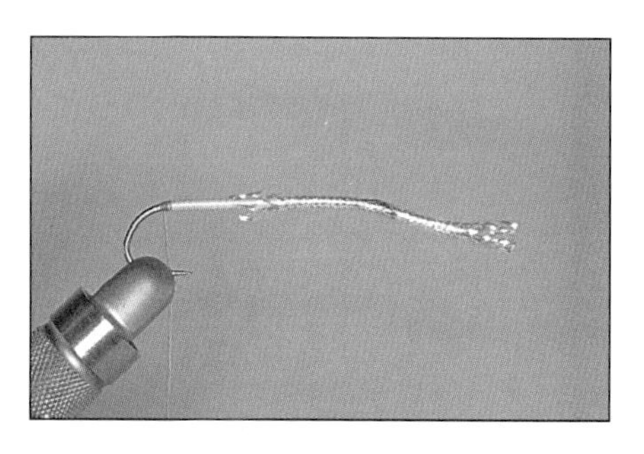

***Running the tubing onto the hook; step 2.***

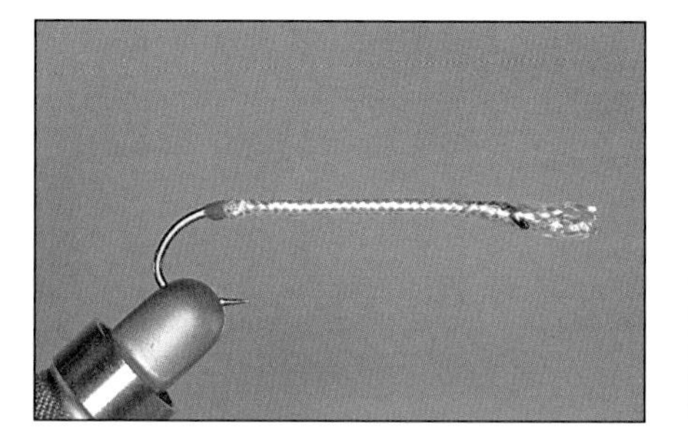

***Tubing secured at rear; step 3.***

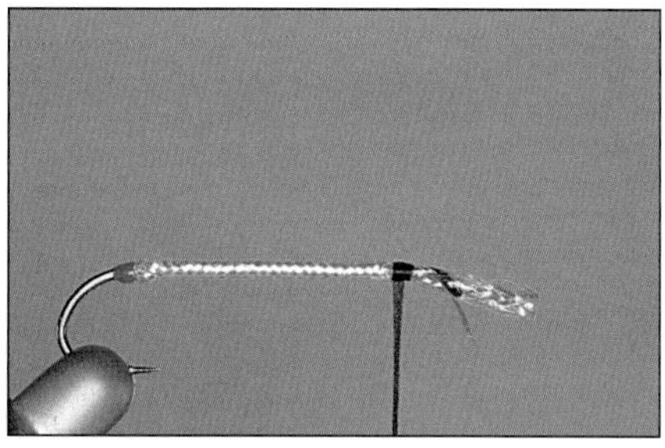

***Black stretch nylon tied on at the front, tubing secured and trimmed; step 4.***

5. Using Ghost Fiber or a similar soft, hairlike material—as described—tie on the first layer of the "wing." Shroud the material down around the sides a little, so that the body and tag are veiled. Trim off the excess. Wrap a base for the next layer.
6. Tie on a small but plainly visible bunch of black hair. (This will simulate the stripe that runs the length of the natural minnow.) Be sure to set it on the wrapped base, so that it doesn't mix in with the white hair. The idea is to maintain separation between the layers. Trim the butt ends on a slope. Or, you may wish to precut the hair to length before tying it on, so you don't have to worry about nicking the Nylon Stretch when you're trimming.

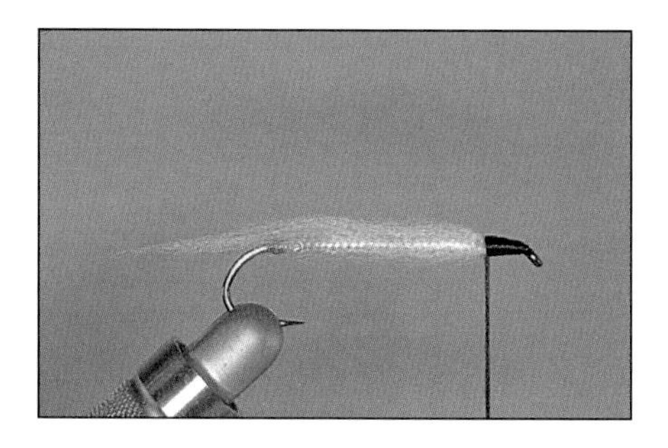

***White material shrouded over body; step 5.***

***Wing layers in place; steps 6-7.***

7. Tie on the top layer in the same manner. Note that it's considerably thicker than the other two layers were. Precut or trim later, as you prefer. Finish off the head, shaping it and building it up with some well-placed wraps. Make your whip-finish back near the rear of the head to avoid having the nylon slide down the front slope.

### *About the Mylar Sheet Eye and the Epoxy Head*

I'll introduce you now to a relatively new and very high-tech item, *Mylar eyes.* They come die-cut, in sheets, and have stickum on the backs. They also come in a variety of colors and sizes.

They're actually a spin-off from the lure-making business, but they certainly make super eyes on streamers, and with such ease! Here's the process:

1. Stick a gold-with-black-pupil Mylar eye of the appropriate size on each side of the head.

*Applying and epoxying the Mylar eye.*

*The completed Dick's Dace.*

2. Mix a little clear epoxy and apply it to the top of the head with a toothpick. Smooth it over the entire head area. As little globules form beneath the head, remove them with the toothpick. Set the fly aside and allow the epoxy to thoroughly set up.

This makes a most attractive, durable, and "fishy" little streamer. The beauty of using the stretch nylon as thread is the speed and simplicity it brings to this school of fly tying. The black does double duty: It secures the layers of hair while also building up the head to the requisite size for mounting the eyes. It wraps very smoothly, which makes it easy to stick on the eyes before you apply the epoxy.

From a practical standpoint, I like to tie these flies in bunches of a half-dozen. This way, I can coat all of the heads with one batch of epoxy. I recommend the slower-acting epoxy; the 5-minute variety sets up so quickly that you may not be able to coat more than a few heads before it becomes too viscous to work with. In any event, you'll need to work quickly during this process.

## The Clouser Minnow

The next streamer represents a departure from conventional design. It has become enormously popular in recent years, and with good reason, as it's a fish-catcher par excellence. It's tied in myriad sizes and color combinations for both fresh- and saltwater fly fishing.

The design is that of Mr. Bob Clouser, a guide, outfitter, and fly-shop operator based in Middletown, Pennsylvania, which is near Harrisburg, on the Susquehanna River. His reputation as a premier smallmouth bass guide on his home river is well deserved. He can also show you how to catch a tiger muskie on a streamer, and—would you believe—catfish on dry flies!

Besides the essential methodology, tying the Clouser Minnow will teach you yet another method for making eyes—one that, at the same time, weights the fly. It will also introduce you to another category of contemporary fly-tying materials, specifically, glitz.

### *About Dumbbell Eyes*

There's no mystery about how these handy items got their name: They look very much like the things the iron-pumpers use in the gym. They come in a variety of sizes, which of course translates to weights. They're also available with shiny finishes, such as gold and silver, as well as the basic leaden color.

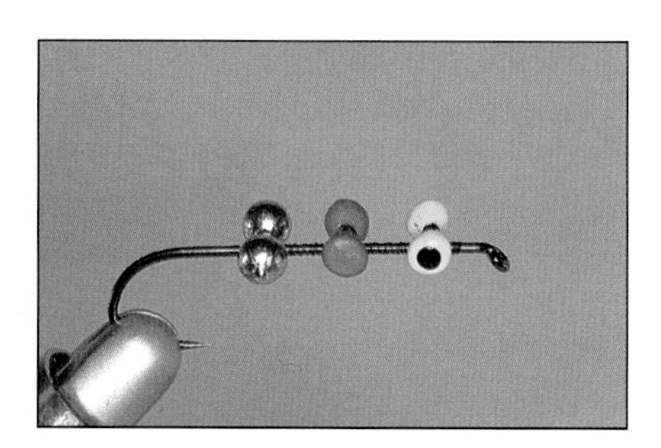

***From left to right: bead chain, plain dumbbell, dumbbell with painted eyes.***

Dumbbell eyes can be left as-is or painted for more realism. If you intend to paint them, don't buy the chromed or plated types, as paint doesn't adhere to those finishes. Before painting the lead eyes, you might want to boil them for a few minutes to remove the residue of casting resin. Soaking them for half an hour in vinegar also removes the resin.

Handling these eyes while applying paint is tricky. You'll want to paint them in bunches. Here are two methods:

1. Insert the eyes between the teeth of a common pocket comb, and apply the paint with a small brush.
2. String the eyes on a piece of fine copper wire by making twists around the "waist" of each eye. Tack a piece of newspaper to the wall in your basement or garage. Make a loop in one end of the wire, and hang the string of eyes from it. Spray them with two coats of a paint designed for adhering to metal, allowing drying time between coats. You can also brush-paint them, using this technique.

Yellow is the most popular color for painting these eyes, but white, green, orange, red, and various metallics are also used. After the paint has dried, you can apply pupils dot-fashion, as I described earlier. You can do this before or after the fly is tied.

### *About Bead-Chain Eyes*

Before the advent of dumbbell eyes, fly tyers used pairs of beads from ordinary electric-light pulls to obtain a similar effect. These can be purchased in hardware stores as well as most fly shops. They come in gold, silver, and brass finishes and in several sizes. You simply cut them into pairs with a wire-cutter, and tie them on by the short section of wire that joins the two beads.

### *About Glitz*

This is not the name of a specific product; it's a catchall phrase that applies generally to the myriad shiny materials in the fly-tying marketplace. The list is a long one, and it's pretty much up to you to select the glitz that's best for the application at hand.

Here, you'll be using a product called Krystal Flash. There are several similar products on the market. It comes in the usual wide array of colors and is used as a highlighter. With the exception of flies for Pacific salmon—which seem to be voracious glitz-eaters—this sort of stuff should be used sparingly. A hint of flash can attract a fish; too much can send the alarmed creature streaking for cover.

Besides streamer wings, this type of material can be used to highlight Woolly Buggers. If you wish, you can tie in a few strands on each side of the marabou tail.

### *About Synthetic Hair*

While we're staying with bucktail for this particular Clouser, many are now tied with synthetic hairs of various types. As with glitz, synthetic hairs have proliferated to the point where they defy description. There's such a wide variety of textures, lengths, and other characteristics that there's virtually no limit to what the creative tyer can fashion out of synthetics. The saltwater-tying contingent has been particularly innovative in using synthetic hairs.

The main thing to keep in mind about synthetic hairs is the range of variation in texture, flexibility, and compressiveness. Some are soft and pliant; others have the consistency of toothbrush bristles. All have their place within the ever-expanding realm of fly tying and fishing. The trick is to choose the type that suits the application at hand.

A particularly useful type, which I've been using for many years, is known generically as *craft fur*. You'll find it in crafts shops as well as fly shops. It's a translucent sort of material, somewhat reminiscent of natural polar bear. However, it's soft and pliable and much easier to work with than most synthetics. Its main limitation is its length, which is adequate for most freshwater flies but less than what's required for larger saltwater patterns. Many colors are available.

In fly shops, you probably won't see craft fur under that name. Fly-tying materials suppliers repackage this stuff under such names as Fish Fuzz and Fly Fur. It's usually a bit more expensive to buy it this way; however, there is an advantage: The repackaged material is better suited to fly tying, in terms of length and texture, than some of the craft fur sold in the hobby shops.

***The Dressing*** Clouser Minnow

*Hook:* Daiichi #1710 or #1720; 2X- or 3X-long nymph or comparable.

*Hook Size:* Medium to large.

*Thread:* 6/0 Uni-Thread or, for larger sizes, 3/0 Uni-Thread or comparable; for this fly, you'll need white and black.

*Head:* Dumbbell eyes.

*Lower Wing:* White bucktail or synthetic hair, if you wish to try it.

*Highlight:* Pale-colored Krystal Flash or similar material.

*Upper Wing:* Black bucktail or synthetic hair.

**Note:** The dumbbell eyes will make this fly ride upside-down in the water. Thus, while the white material is tied on top of the hook, it actually becomes the belly.

### Tying Steps

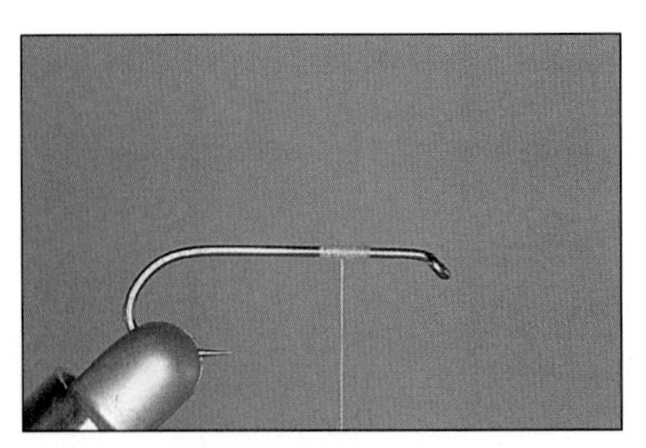

***Tie-on position and thread base; step 1.***

1. Tie on, using the white thread, and wrap a thread base, keeping in mind that the dumbbell will be mounted about 1/3 of the shank length rearward of the eye. Allow sufficient space to tie down the materials ahead of the dumbbell eyes.
2. The dumbbell is mounted on top of the thread base. Hold it in the position shown and take three wraps around its "waist," passing the thread across from back to front. If you want realistic eyes, use a dumbbell that's already been painted yellow (or whatever you want the base color to be).
3. Hold onto the dumbbell from the near side and take three front-to-back wraps of thread, crisscrossing the first three. This brings the dumbbell into position: squared off on top of the hook.
4. Put a generous drop of Zap-A-Gap on the wrappings, then continue to make crisscross wraps until the dumbbell is secure.
5. Cut off a small bunch of white or pale-colored hair, and clean out all shorties and

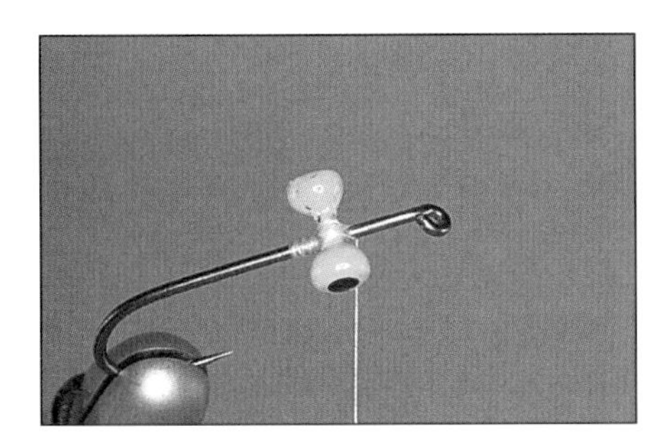

*Dumbbell figure-eighted on; steps 2-4.*

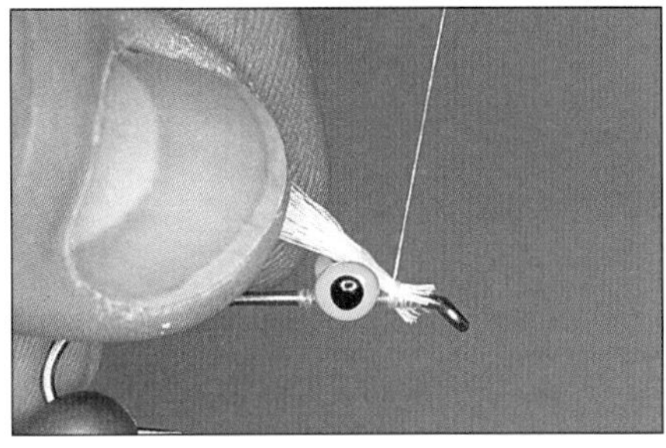

*Tying on white hair; step 5.*

underfur. Pretrim the butt end of the bunch. Tie it on in front of the dumbbell, tilting the bunch slightly upward; this offsets the hair a bit at the front end and makes it easier to form a nicely tapered head. Lock down the bunch with some very firm wraps.

6. Pass the thread underneath, and take a series of wraps around the hair rearward of the dumbbell, as shown. Then cross back underneath, and switch to black thread.
7. Turn the hook upside-down in the vise. Tie on a small bunch of Krystal Flash—perhaps eight fibers in all. Trim off the butts, and cut the strands to the proper length at the rear.

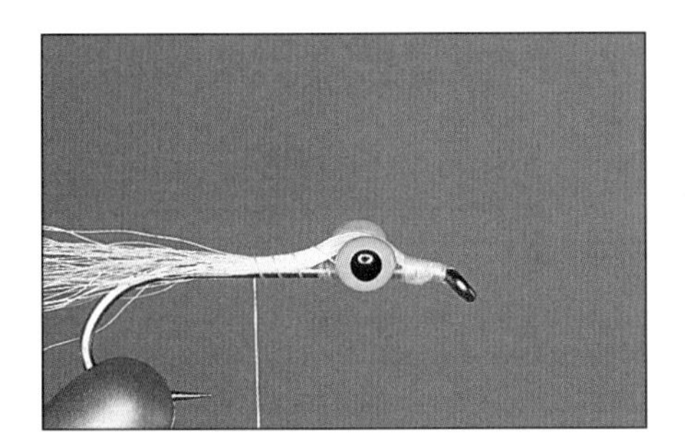

*Securing hair rearward of dumbbell; step 6.*

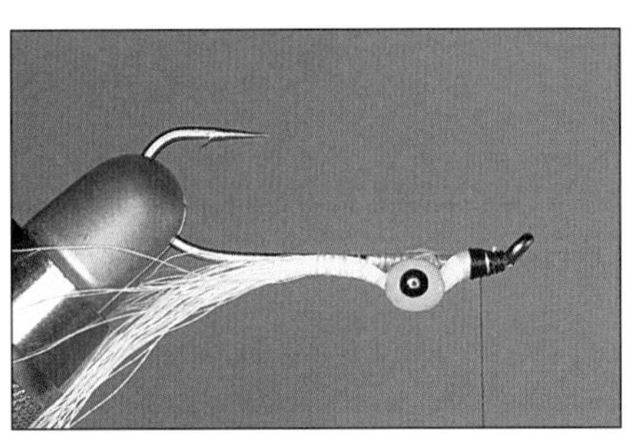

*Hook inverted, black thread tied on; steps 6-7.*

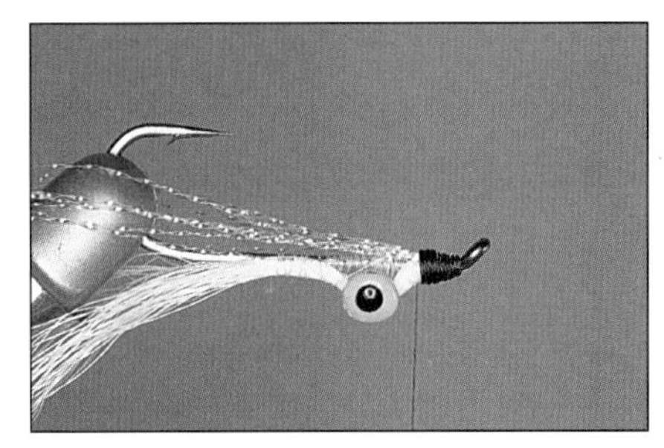

*Krystal Flash tied on; step 7.*

8. Prepare a bunch of black bucktail, or whatever hair you've chosen, as you did the white. It should be somewhat thicker than the white, however—perhaps twice the amount. Pretrim it and tie it on, again tilting the bundle to offset the butt ends. Secure with firm housekeeping wraps and whip-finish.

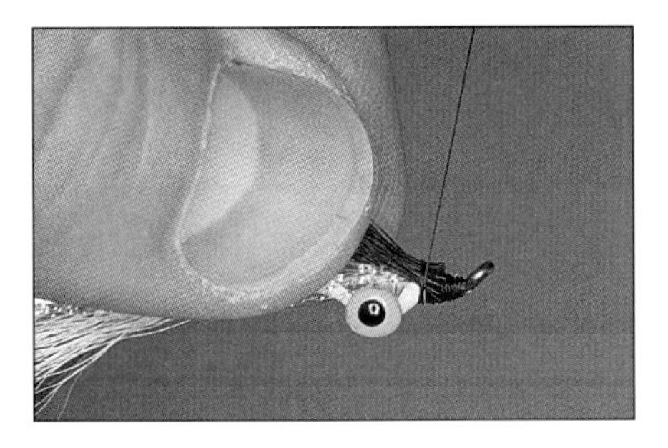

*Black hair being tied on; step 8.*

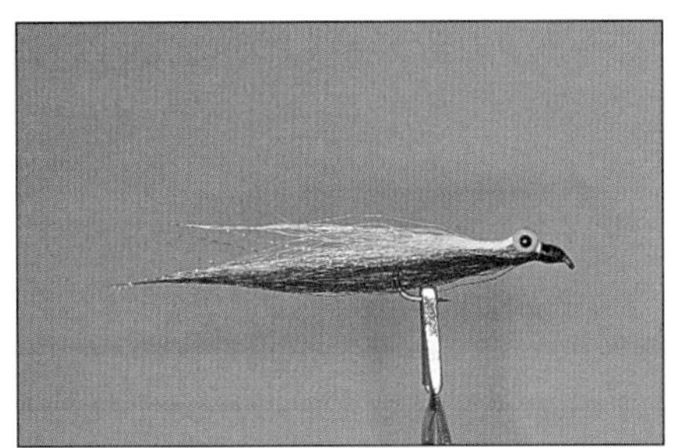

*The completed Clouser Minnow.*

9. It's a good idea to Zap-A-Gap the spot between the eyes where the belly hair crosses over, as well as the wraps that secure it behind the dumbbell. In fact, there's no reason that you couldn't simply use Zap-A-Gap as head cement and coat all the wrappings with it. If you want a truly indestructible head-eye assembly, cover everything with epoxy.

This particular dressing produces a fly that looks like a baby catfish, or perhaps a tiny elver. As I stated, there's virtually no limit to the combinations you can put together using this basic design. As you've probably gathered, this fly fishes very deep; in fact, it's also known as the Clouser Deep Minnow. When retrieved, it has a deadly jigging action. The down-eye hook, when inverted, becomes an up-eye, which helps you avoid snags when you're fishing the fly along the bottom.

## The Light Spruce Fly

The next fly, the beautiful Light Spruce, started out as a pattern for the Pacific Northwest but has since gained popularity elsewhere. I can personally attest that the brook trout in my region find it very appealing in smaller sizes. The dressing introduces you to another type of peacock and provides you more practice both setting in place the feather-type streamer wing and tying the folded-and-wrapped collar-style hackle.

Two types of peacock go into this dressing: The front portion of the body is made of the regular herl you used for the Prince Nymph, while the tail calls for *peacock sword.* When you see such a feather you'll have no problem understanding where the name came from. These saberlike feathers are found along the flanks of the bird. The herl is iridescent and sometimes has bluish and bronze highlights. Peacock sword is a fairly common ingredient in older patterns, particularly traditional salmon flies. One dressing, the Alexandra, features an entire wing made of peacock sword.

The wing and hackle for this fly are made of a uniquely marked feather called *badger;* this, I would guess, because the dark center stripe looks like the one on a badger's face. There are various shades of badger; here, you'll want the lighter shade, which is known as silver badger. For a while it was in short supply, but it's made a comeback thanks to the popularity of fly tying in general and of this pattern in particular.

The appearance of your finished fly will very much depend on how well you match up the wing feathers. This is another argument for buying them as a pelt, rather than strung: It's often next to impossible to find well-matched feathers in a strung bundle, especially when you have to match up two sets with opposing curvature for the back and front wings.

I should also mention that this fly's original dressing called for red floss for the rear portion of the body. I've substituted Uni Nylon Stretch, simply because I love its Chinese red shade and think it better suits this application. You'll need a spare bobbin in which to mount this material, because you'll use it to tie on the tails as well as to form the rear part of the body, as was done on Dick's Dace.

**The Dressing** Light Spruce Fly

*Hook:* Daiichi #2340 or #2370 streamer or comparable.

*Hook Size:* For this exercise, size 2 or 4.

*Thread:* 8/0 Uni-Thread or comparable; black.

*Tail:* Small bunch of peacock sword herls.

*Rear Body:* Uni Nylon Stretch in Chinese red or similar material.

*Front Body:* Peacock herl.

*Wing:* Four silver badger hackles.

*Collar:* Silver badger.

## Tying Steps

1. The proportions of the body are about 60 percent red material and 40 percent peacock. If you're using stretch nylon for dual purposes, mount it in a bobbin and tie it on thread style about 40 percent of the shank length rearward of the eye. Trim the tag and wrap almost to the bend. If you're using the conventional thread method shown here, follow the same procedure with the black thread.

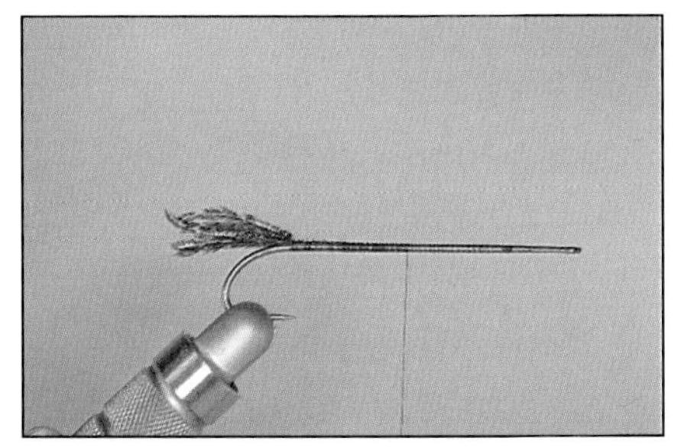

***Tail tied on, thread positioned to tie in body material; steps 1-2.***

2. Tie on a short tail, using the body material. Taper-cut the tail butts and cover them as neatly as possible. If you're using stretch nylon, wrap to the tie-in point, then back to the base of the tail, then forward again, forming the rear portion of the body. When you reach the tie-in point, whip-finish the stretch nylon and trim off the tag. Tie on with black thread, covering the tie-off of the body material for added security.
3. If you're using the conventional thread method, after tying on the tail and trimming the butts, wrap forward, forming a smooth underbody. Just beyond the halfway point, tie in a piece of stretch nylon or similar material, and wrap it to the rear and forward again, thus forming the rear segment of the body. Trim off the excess.
4. Tie in a bunch of peacock fronds by the tip ends, make a thread loop, cut off one side, and twist it in with the herl, then wrap and twist forward. This is the same procedure you used on the Prince Nymph. If you're tying a large-sized streamer, use thick-herled peacock.
5. A feather wing doesn't require you to allow as much head space as a hair wing does, but do leave enough for the collar to follow. Tie off the herl and wrap a smooth thread base.

***Rear portion of body in place; step 3.***

***Herl portion of body in place; steps 4-5.***

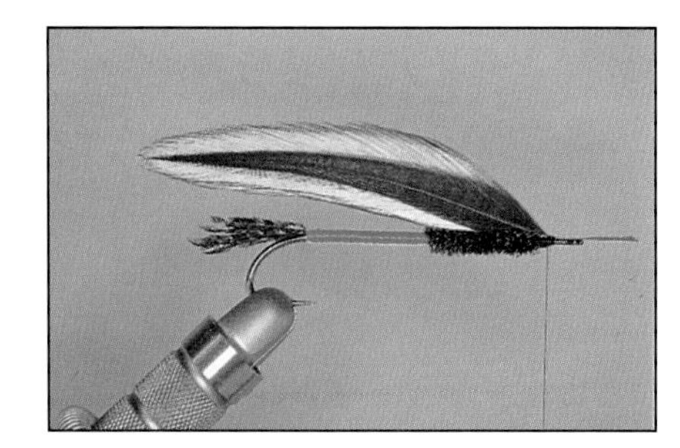

***Wing in place; step 6.***

6. Select four well-matched badger hackles. If they're all straight, fine. If they have significant curvature, you'll need two fronts and two rears. Cup them together, and strip back the quills to length. Go slowly, because you can't put the barbs back on once you've removed them; study the proportions in the photo. Be sure the tips of all four feathers are even at all times. When you've attained the proper size, remove just a few more barbs from the bottom of the quill to allow for the front of the body.
7. Quills may be precut, just like hair, and I recommend this. In fact, you can obtain a smoother head by precutting the wing butts on a slope with a razor blade. Set the

wing assembly in place and tie it on with a series of increasingly firm wraps. Lock them in well. Then take a few housekeeping wraps, as appropriate, ending up with the thread at the front of the wing.

8. Select a large badger hackle—one at least the size of one of the wing feathers. Ideally, it should be taken from the center of a cape, so that the barbs are of equal length on both sides of the quill. Strip off the waste material at the butt end. Then make like you're going to barb-gather—but instead of pulling off material, stroke the barbs down on the respective sides of the quill from which they grow. No crossovers. As I stated earlier, hackle pliers make a great handle when you're performing this procedure.
9. Tie in the hackle near the tip end, at the point where the barbs are of sufficient length to make a nice collar. Secure it well, then trim off the excess and cover with a few housekeeping wraps.
10. Wrap the hackle per the instructions in chapters 2, 3, and 4, stroking back the barbs as you go and making the quill the leading edge. This sort of wrap works better without hackle pliers. Complete just behind the eye, trim off the quill, secure, whip-finish, and you have a streamer fly of great attractiveness to fish and angler alike.

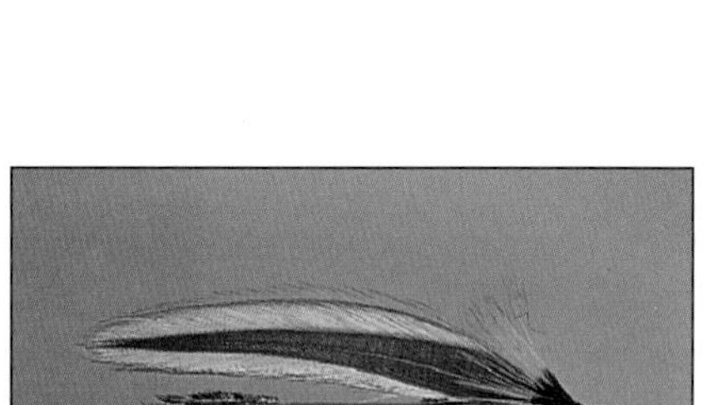

***Hackle collar wrapped, completing the Light Spruce Fly.***

One afterthought. If you feel that you need a fuller hackle, you can use two feathers at once. Simply follow all of the above instructions while treating two feathers as one. They must remain in perfect alignment throughout the process, with the quills lying together.

For our final streamer exercise, we'll tie a Matuka. There are a number of variations of this style of streamer, some of which call for feather wings and others of which use strips of rabbit pelt. We'll tie the latter style, keeping in mind that the feather version is tied with the same method.

There's also a variety of techniques for making heads, collars, and "front ends" on these flies. Here, we'll make a spun-hair collar/head assembly. This is a bit of a challenge for the beginning tyer, and you may want to hold off until you've become fairly competent at fly tying in general. The reward for learning this technique is that you can then tie, among other things, the redoubtable Muddler Minnow—perhaps the greatest all-around "meat" fly ever.

I like to think of this fly's tying procedure as being divided into two distinct segments. The first is the tying of the components up to and including the "wings." The second is the hair collar/head. There are two essential factors in the relationship between the segments: isolation and integration. *Isolation* means it's important that the first segment be successfully and securely completed before the second is begun. *Integration* means that the two segments must fit together properly and complement each other.

### *About Hair for Spinning*

A number of hairs lend themselves to being spun and trimmed: elk, antelope, caribou, and various species of deer. Some are coarse and best suited for bass bugs, hair mice, hair frogs, and such work. Antelope hairs are particularly adaptable to bass bugs, as they're quite thick and have a honeycombed structure inside for optimal flotation. Caribou is fine and soft and works well for smaller flies such as the Irresistible and Rat-Faced MacDougal. Elk hairs from certain sectors of the animal's body spin okay, but elk is more commonly used for hair wings, and such.

Usually, we obtain our collar/head hair from one sort of deer or another. Given the wide variety now available, it's possible to find the right stuff for any sized fly. The criteria are both texture and length: The length of the hair itself, and how long a collar is desired, dictate the point at which the thread will intersect the hair bunch.

What we're creating on this fly is a trimmed head, fashioned out of the butts of the hairs that form the collar. The collar shouldn't flare too much. The hairs for the head should crimp to approximately 90 degrees to facilitate trimming. A single bunch of hairs will accommodate both head and collar if it's "hard" near the tip ends and begins to soften just where the thread wraps used to spin the collar into place intersect the bunch. I hope this is clear.

On smaller flies, the collar and head are formed out of a single bunch. On medium-sized flies, the collar and part of the head are formed out of one bunch, and a supplementary bunch is used to fill out the head. These bunches will be about the diameter of a pencil, give or take a little. In both cases, we employ the spinning technique, where the hair is induced to deploy 360 degrees around the hook. You'll see how this works in the tying sequence.

On large flies, two bunches are required for the collar, plus one or two additional bunches for the head. Here we get into a variation of technique, as we want the two collar bunches to deploy only 180 degrees—one on top, one on the bottom. This is known as *stacking.* It's the same as spinning, with one exception. This will be covered in a photographic sequence.

There's one bone of contention that I'd like to address forthwith. Traditionally, descriptions of hair-spinning advocated applying the hair to a bare hook shank. I agree that it does spin a little more readily that way. However, most contemporary hair-spinners work over a thin, smooth thread base, and I find that this suits me better, too. My heads and bodies come out stronger and more durable. Suit yourself.

The *type of thread* you choose also relates to size of your fly and to the texture and amount of hair. For small work, I find 6/0 Uni-Thread highly suitable. For larger flies, I use 3/0 Uni-Thread, Dynacord, or Monocord. For all its fantastic strength, I don't care for Kevlar; it just doesn't behave well.

**The Dressing** Rabbit Matuka

*Hook:* Heavy wire; 4X or streamer length.

*Thread:* Same as body; then brown 6/0 Uni-Thread or comparable.

*Body:* White Uni Nylon Stretch or comparable, mounted in a bobbin.

*Ribbing:* Oval gold tinsel or wire, also mounted in a bobbin (see note).

*Wing:* Natural-color strip of tanned rabbit pelt.

*Collar/Head:* Deer hair, per prior notes.

**Note:** Mounting the ribbing tinsel in a bobbin frees up the hands for working on the wing which securing it in place.

### Tying Steps

1. With the body material mounted in a bobbin, tie on about 1/4 shank length rearward from the eye and wrap to the halfway point.
2. Tie in the tip end of the ribbing tinsel, and wrap to the rear, securing it.

3. Wrap up and down the hook two or three times, building up the body to the desired diameter; then whip-finish and trim off the body material. The tinsel is hanging off to the rear, suspended by the weight of the bobbin.
4. Tie on with the brown thread, make a thread base, and tie in the strip of rabbit atop the hook. Note that plenty of space is retained up front for the collar/head assembly.

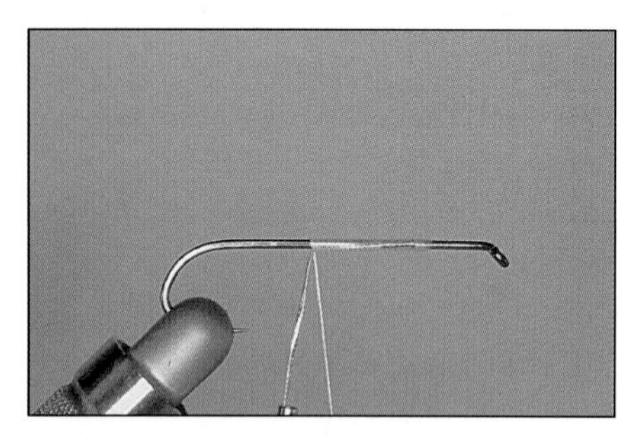

***Tying in the tinsel with the body material. Both are mounted in a bobbin; steps 1-2.***

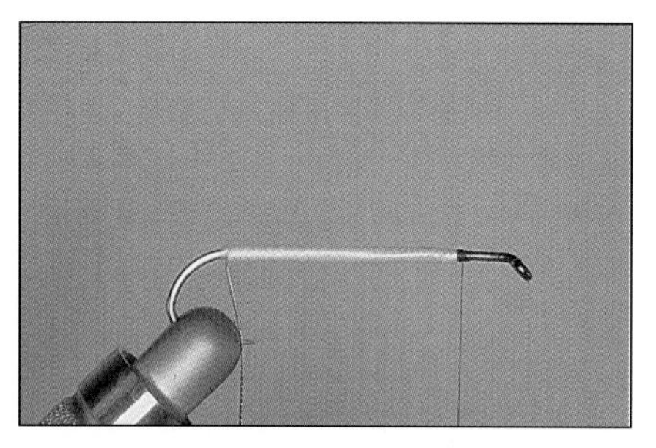

***Body formed, thread tied on; step 4.***

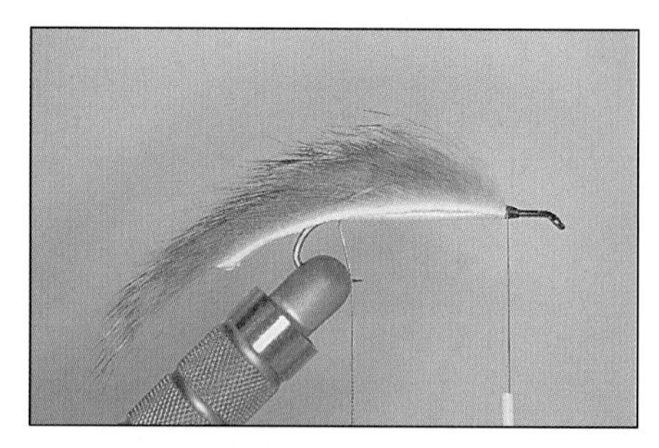

***Rabbit strip tied on; step 4.***

5. At the rear, separate the rabbit fur to expose the hide exactly where the tinsel is hanging. A toothpick or something similar is helpful in doing this. Then pass the tinsel through the notch, as shown.
6. Snug up the rabbit strip by pulling from the rear. Then continue forward, separating the fur at regular points, so that the ribbing goes on evenly. Keep the rabbit strip taut by pulling from the rear periodically.

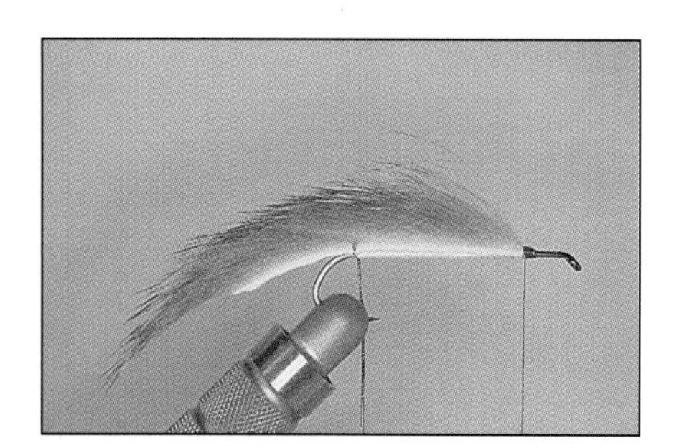

***First turn of tinsel at rear; step 5.***

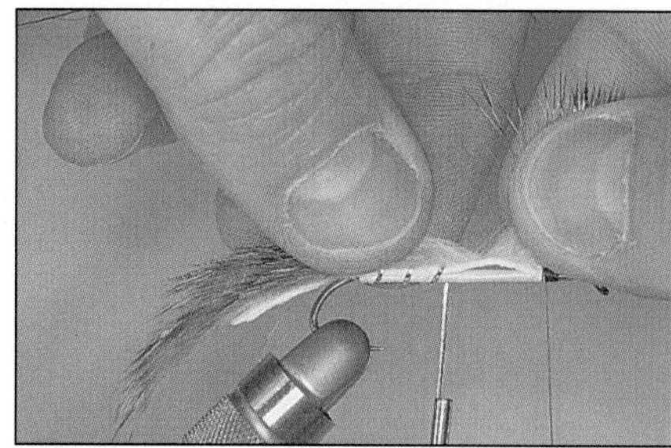

***Ribbing and securing; step 6.***

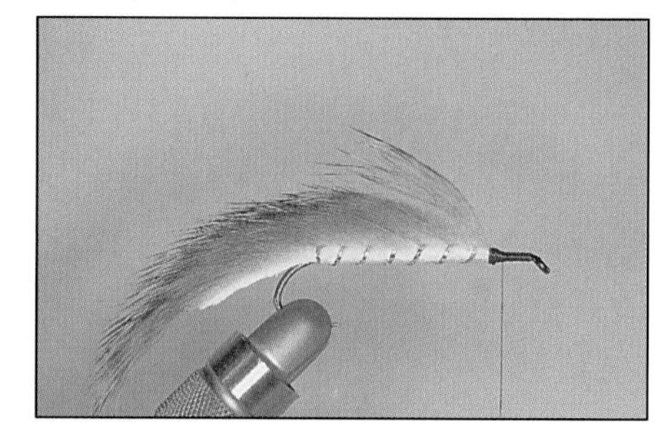

***Ribbing completed, thread positioned for collar/head; step 7.***

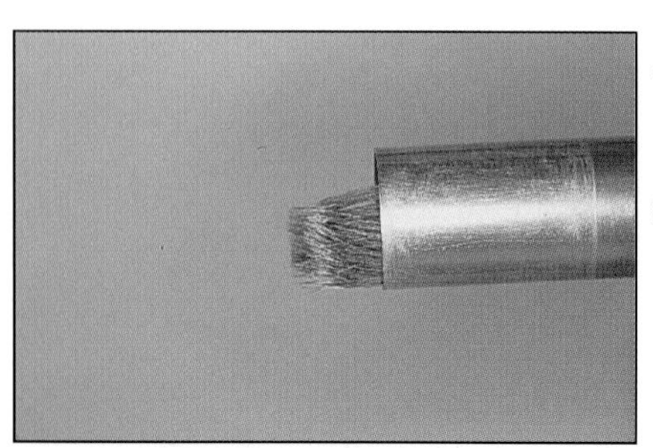

***A bunch of evened hair; step 8.***

7. When you reach the front end of the rabbit strip, tie off the tinsel. Then wrap neatly to the eye and back. The thread should be hanging at the front of the rabbit strip.
8. Cut off a generous bunch of deer hair. With your fingers or a fine-toothed comb, clean out all the underfur and shorties, then even up the tips in a hair stacker. While holding the bunch by the tip end, cut the butts so that they're slightly longer than what you'll require to form the head. This keeps the butts, which will be trimmed, from getting all mixed up with tips, which won't be trimmed. It also makes it easy to pass the thread forward through the hair.

9. Be sure the thread is tight—that is, not spread into strands. If necessary, spin the bobbin clockwise. Hold the bunch against the far side of the hook, as shown, positioning the tips for collar length you desire. Take one encircling thread wrap under moderate tension; then a second, precisely over the first, with more tension; then a third. As you're about to complete the third wrap, let go of the hair. It should spin 360 degrees, fully encircling the shank.

***Hair butts trimmed, bunch positioned for spinning; steps 8-9.***

***First two wraps in place; step 9.***

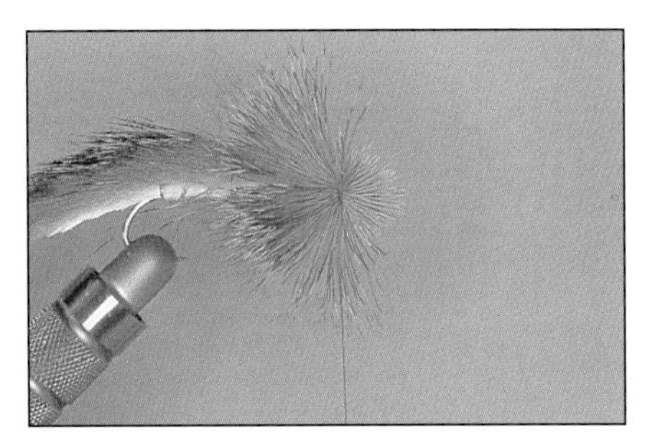

***Third wrap applied and tightened, hair spun; step 9.***

10. Take a couple more firm wraps in the same spot. Then stroke the butts upright and spiral the thread through them to the front of the bunch.
11. Pack the hair tightly with either your fingers or a packing tool.
12. If you have enough hair for the head, tie off. If not, cut off another bunch and clean it, but this time use just the butts or midsection (whatever looks best to you), cutting off the tips. This eliminates having to figure out what's the collar and what should be trimmed into the head. Spin it on; pack and secure it, just as you did the first bunch; then tie off (see note following).

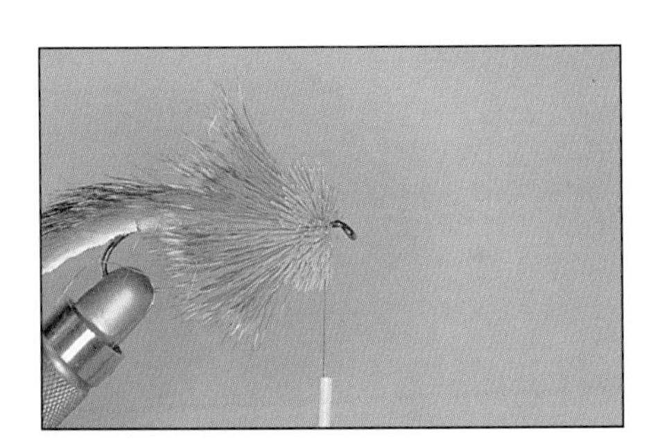

***Hair butts stood up, thread advanced to front; step 10.***

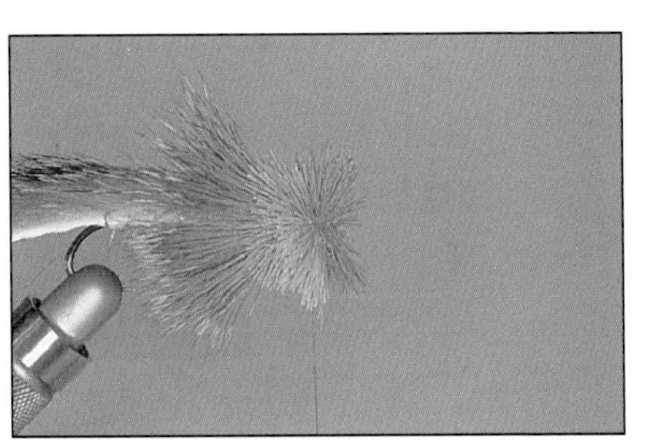

***A second bunch of hair added, with tips trimmed. Compact the hair with either your fingers or a packing tool (shown here), whether supplementing the hair or not; steps 11-12.***

13. Trim to whatever shape you like. I prefer to cut the bottom flat, collar and all, then shape the top so that it's wider than it is high—like a sculpin. True experts often use a double-edged razor blade on the bottom, trimming with one stroke. Be my guest, but remember: If you nick the thread, you're out of business. For shaping the head, I use curved scissors with tiny serrations; they grip the hair better. Work with the points toward the rear, so as not to cause the hairs to lift, which would result in uneven cutting. Don't make mincey little snips; use long, smooth, front-to-back

cuts. Work back to where the collar begins. In fact, if the collar is well filled out, you can trim back into the front hairs a little, thus attaining a smooth, well-integrated transition.

*Trimming, top and bottom; step 13.*

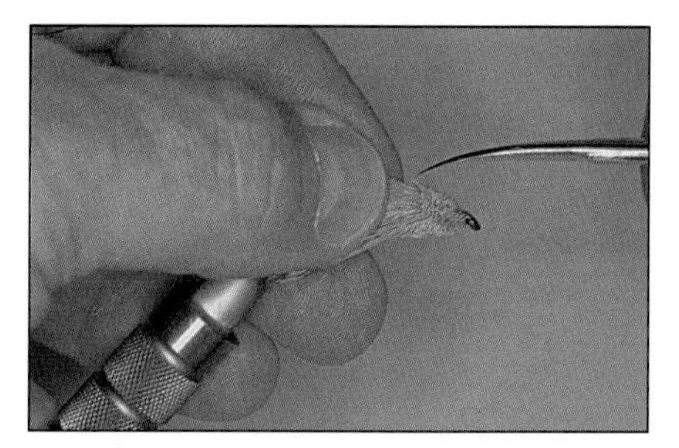

*Fine-tuning; step 14.*

14. Invariably, a few butts will be buried in with the collar. To expose them, simply pinch the collar flat against the wings from the sides. The butts will stick out, and you can easily trim them off. When you're all done, apply two coats of head lacquer to the whip-finish knot.

*The completed Rabbit Matuka.*

*The two steps in stacking hair.*

**Note:** For those who have problems creating a whip-finish with all that hair up front, I have two suggestions: use a hackle guard, or use a Matarelli whip-finishing tool. The latter is a most handy instrument: Once the thread is started you can hang the bobbin in the materials clip, or off the rear of the vise, and make the whip-finish knot one-handed, while using your other hand to hold the hair back. Very neat.

Now for the stacking procedure, which I alluded to earlier. It's called for when you need a collar so large and full that one bunch of hair won't suffice. Here's the drill:

1. Some tyers find it easier to turn the hook upside-down in the vise for the first stack, then restore it to normal position for the second. Suit yourself. Other than this, start exactly as you would for spinning on a collar, except that the hair bunch is held precisely atop the hook. Make your thread wraps, but instead of letting the hair spin, hold it in place while you complete the third wrap. Then add a few more tight wraps, still holding the hair in place. This results in a 180-degree half-collar.

2. Tie in a second stacked bunch that's a mirror image of the first. Stroke the butts up straight, work the thread to the front, and pack.
3. Now revert to the spinning technique, adding as much more hair as is required to complete the head, and finish the fly per previous instructions.

### *About Rabbit Pelt Strips*

Commonly referred to as *zonker strips,* these consist of narrow strips cut from a tanned rabbit hide. Many suppliers sell them precut, which, while more costly, is definitely a convenience. They can then be cut to length for the fly at hand.

Rabbit is sold in natural shades as well as in every conceivable dyed color. Quality will vary, so be sure to inspect the pelts you buy to ascertain that they have good coats. Also, watch out for ones that have thick skins, which some do. These are harder to work with and don't lend themselves as well to fly tying as do the thinner-skinned ones.

If you choose to buy whole pelts, you'll be faced with the task of cutting the strips yourself. This isn't really difficult, if you have four hands (I'm being facetious, of course). The job is easier with two people, but it can be done by an individual with an assist from a couple of makeshift tools. Here's how I cut my rabbit strips:

1. Take a board about 2 feet in length and at least 6 inches in width—any old board will do—and drive a nail in the center near the top end.
2. Get yourself one of those enormous clips sold in the supermarket for closing potato-chip bags. Clamp it onto the front of the rabbit pelt, so that the pelt will lie lengthwise.
3. Hook the handle of the clip over the nail, with the hide side of the pelt facing outward.
4. With a very sharp razor blade or craft knife, cut your strips, while holding the pelt tight from the bottom. It takes a bit of care and practice to get them to come out even. They should be quite narrow: $\frac{1}{8}$ inch or so.

***A streamer and a minnow.***

***A dry fly natural and imitation.***

◂ S I X ▸

# *Dry Flies*

Dry-fly fishing is considered by many to be the ultimate form of angling. Some fly fishers limit themselves to this method exclusively, which is their choice, albeit not the most practical one. I fly fish by whatever means works best in a particular situation, but I'll readily admit that whenever I can fish dry and have a reasonable chance of success, that's what I'll be doing.

Tying dry flies is also a lot of fun. Few pleasures in life compare with those of composing the perfect wood duck wing or wrapping the perfect hackle. And dry flies aren't all that difficult; if you've gone through all of the exercises in this book to this point, you've already completed procedures that are trickier than dry-fly work.

I believe that, along with technique, the critical factor in dry-fly tying is understanding materials and using those that best suit the application at hand. Preeminent among these is hackle; let's learn about that commodity first off.

## Dry-Fly Hackle Terms

***Barbs*** Just as with wet-fly hackle, this term refers to the fibers that protrude when the feather is wrapped, thus forming the hackle around the thorax area of the fly.

***Barb Count*** This refers to the number and proximity of barbs on a given length of quill.

***Cape, Neck*** These terms are synonymous. They refer to a pelt taken from that part of a chicken.

***Saddle*** This is a pelt taken from the back of a chicken.

***Sweet Spot*** This is the part of the feather that contains prime dry-fly hackle. Further definition will follow.

***Size*** In this context, size refers to the size of hook a feather will properly accommodate. Feathers are referred to as size 12 or 14 or whatever. This is determined by barb length.

***Web*** This refers to the softer parts of barbs adjacent to the quill.

***Spade Hackle*** These are the larger feathers found along the edge of a cape. They're also known as *throat hackle.* Their long, stiff barbs make them desirable for dry-fly tailing.

## *Judging Quality*

Hackles for floating flies must possess certain qualities. Here's a list of the critical ones:

- A fine, flexible quill.
- A high barb count; that is, lots of barbs per wrap.
- Strong barbs, all the way out to the tips.
- Uniform length of barbs throughout the usable length of the feather.
- Straightness of barbs—or at least relative straightness.
- Absence of significant web in the center.
- Rich coloration; an attractive "sheen."

The first four items most directly affect quality and are rather interrelated—and thus difficult to rank in absolute order of importance. However, I do feel that the quill is perhaps the most important consideration. Given a very narrow quill that wraps true, you can compensate for other attributes that are less than optimal. For example, if the barb count isn't very high, you can use more turns and pack the wraps more densely. This will also compensate, to a degree, for barbs that aren't quite as strong as you might wish. The results are pleasurable tying and dry flies that perform well astream.

## *Grading*

This term applies to the informed analysis of hackle—individual feathers as well as complete pelts. Both capes and saddles are graded by the grower before being sent to the retail store; this determines how much the shop—and subsequently, the customer—pays. Most growers grade on a 1-2-3 scale, with 1 the highest. Occasionally, you may see some "off-grade" pelts offered for sale at reduced prices. If they come from a reputable grower, look them over carefully; you may be pleasantly surprised and end up with a bargain.

Birds vary considerably from flock to flock, which means that grading becomes somewhat arbitrary. This is why it's important that you, the consumer, learn how to make these assessments for yourself. As a fly tyer, you're in a better position to do this than many of the people in the business. Even so, everyone is essentially in agreement that the quality considerations mentioned above, along with the quantity of material on a given pelt, determine grade.

Is color a legitimate factor in grading? The answer is both yes and no. Here's what I mean: Certain colors, or strains, of chickens don't produce as good a grade of hackle as others. This is simply a matter of genetics. These strains can be improved, to an extent, by selective breeding, optimal care, and such, but their hackle will still fall somewhat short of that from a genetically superior strain. In other words, certain colors, such as gray or dun, don't directly compare in quality to certain others, such as ginger or cream. Thus, grading must be done on a relative basis, within color groups.

## *Capes or Saddles?*

While saddles have advanced a couple of light years over the past few decades, capes are still the more popular material, mainly because they come in a wider range of sizes and have additional applications, such as tailing and winging dry flies and, possibly, streamer flies. In addition to the specific quality points in the preceding list, here are a few more attributes to look for when buying capes:

- Lots of feathers *in the size range you want to tie in.*
- Usable feather length—that is, long sweet spots.
- Tailing material (which is getting to be a scarce commodity).
- A well-cleaned cape, free of body fat or grease.

As I mentioned, saddles are coming on strong. The incredible Hoffman saddle pelts from Whiting Farms, Inc. are now generally available in better fly shops and are causing quite a stir. The Hebert roosters, owned by Ted Hebert of Laingsburg, Michigan, are known more for their wonderful cape hackles, but Ted is also producing some fine saddles these days. The Hoffmans tie quite small, whereas Ted's stuff runs larger, so between the two types a wide range of sizes is covered.

Speaking of which, size range is one of the main differences between capes and saddles. A good cape will have seven, eight, or perhaps even nine sizes on it, whereas a saddle rarely possesses more than three and may have only two.

Today, top-grade saddle hackle compares favorably in quality with cape hackle. In some respects, it may be even better; its barb count tends to be high, and its quills are usually thread-fine and eminently wrappable. The sweet spots are quite long; in fact, a top-grade saddle feather is almost all sweet spot. This enables you to dress several flies from a single feather.

### *The Matter of Color*

Much has been made of this topic over the years, and pattern books that list traditional dressings specify all sorts of subtle shades and hues. That's fun for the archivist, but as a fledgling tyer, don't get overly concerned about color. It's very much subordinate to quality, and so long as you're reasonably close, there's no problem.

Perhaps more important than specific color is the tone of the pelt. Healthy birds whose feathers have the quality points I listed above will also have lustrous coats. It matters almost not at all whether the color is natural or obtained by a *properly done* dye job, because modern techniques produce marvelous results without damaging feathers in the least.

Minor variations in color are practically infinite in number. There are also marked, or patterned, feathers, such as grizzly. None of this can be adequately described in words, and I think I can best serve you by referring you to the captioned color phototgraphs of hackles included in this book.

I'll summarize my thoughts on color thusly: If you have four good-quality capes—a medium gray, a straw cream, a medium brown, and a grizzly—you'll be equipped to tie a lot of terrific flies and catch a lot of fish.

Now that you understand some things about hackle, let's use it. Your first dry fly will be a very simple one; in fact, it's made of nothing but hackle! It's called a Bivisible, because it uses two different hackles: one to obtain the overall color and a white one in front to enhance visibility. Of course, you'll be tying it on a dry-fly hook, so let's start by briefly examining these.

### *About Dry-Fly Hooks*

As I mentioned back in chapter 2, dry-fly hooks have—or should have—a number of attributes that differentiate them from other types of hooks. They are:

***Hackle, from top to bottom: cream, straw, ginger, light ginger, cree, medium brown, coachman brown, black.*** Photograph by Tim Savard.

***Hackle, from top to bottom: light gray dun, sandy dun, medium gray dun, barred rusty dun, grizzly dyed iron blue dun, grizzly cape, grizzly saddle, badger.*** Photograph by Tim Savard.

- Light to extralight wire.
- A Model Perfect (semicircular) bend to set the tail into proper position.
- A very small barb and a somewhat shorter point than heavier-wire hooks.
- A moderately turned down eye; an angle of about 30 degrees works well.
- A well-formed eye with no open spot.
- For standard models, a shank-to-gape ratio of 2:1.

Needless to say, metallurgy is much more critical with such fine-wire hooks. There is virtually no margin for error, and too soft or too brittle a wire will result in a hook that either bends or breaks. Even with today's technology, you'll still encounter such hooks from time to time, especially in the lower price ranges.

Fine wire also makes the debarbing process somewhat touchy. The barb on a hook is formed by making a little slice into the wire and raising the cut part. If the slice is too deep, even careful debarbing can cause the hook to break. Some manufacturers use a more brittle wire for dry-fly hooks than for other models in an effort to prevent bending. This is fine, but if it's carried to excess it, too, can contribute to problems with debarbing.

The time to debarb a hook is before you go to the trouble of tying a fly on it. With a pair of flat-jawed pliers, gently pinch down the barb until it either flattens or breaks off, in which case you'll feel it go "pop." There may be a bit of a nub remaining. Leave it alone; you've done all that you can safely do.

For our first dry fly, we'll tie a Bivisible. This is the dry-fly equivalent of the Woolly Worm: simple design, easy to tie, yet effective. It's a particularly good riffle and fast-water fly. The name derives from the white front hackle, which gives the fly high visibility. A number of hackle colors can be used for the body; grizzly and brown are the most common.

A word about hackle sizing. On conventional, winged dry flies, the rule of thumb is that the hackle should be 1½ times the gape of the hook in length. On the water, this will balance the fly with its tail, which is about the length of the hook shank. However, on flies that have *palmered* hackle—hackle that is wrapped from the rear of the hook to the front—better balance is achieved by using a hackle that is only about as long as the gape. This is especially true of Bivisibles, as well as certain types of caddis, stone fly, and terrestrial imitations.

I began tying Bivisibles back around 1960, and I recall having a tough time finding hackles that would accommodate the design. Usually, I had to use two feathers in tandem for the main hackle. Today, a good-to-excellent-quality neck hackle will adequately dress a bivisible. However, if you have a dry-fly-grade saddle, particularly a Hoffman, I suggest using it. Its feather length, quill flexibility, and barb characteristics make it ideal for this sort of work. You'll be able to tie several Bivisibles from one high-grade saddle hackle.

I should mention that the Bivisible may be tied with or without a tail. Here, I've omitted that appendage to make your introduction to dry-fly tying as simple as possible. If you want to add one, make it the same color as the main hackle and the length of the hook shank.

***The Dressing*** Grizzly Bivisible

*Hook:* Daiichi #1180 dry-fly or comparable.

*Hook Size:* For this exercise, size 12 or 14.

*Thread:* 8/0 Uni-Thread or comparable; black.

*Tail (optional):* Same as main hackle.

*Main Hackle (Rear ¾):* Saddle, preferably Hoffman; grizzly, or high-grade cape hackle.

*Front Hackle (Front ¼):* Either saddle or cape; white.

### Tying Steps

1. Tie on about ⅓ of the shank length behind the eye, and make very neat, contiguous wraps to the bend. An aid in making contiguous wraps is to not trim off the tag of thread but instead hold it tightly with your left thumb and forefinger at about a 45-degree angle. This coerces the wraps into neatly aligning. When you're within a few turns of the bend, trim off the tag.
2. Now for sizing the main hackle. I suggest you do this before you remove the feather from the pelt. If you have a hackle gauge, use it. If not, hold the feather near the gape of the hook, so that you can "eyeball" the proportions. Remember, you're looking for barbs that are about equal to the gape in length *in that portion of the feather that will be used to form the hackle,* as barb length can vary throughout the length of a feather. Very high-grade saddle hackle can be a bit deceptive in the sizing process because the extra-strong barbs resist standing out perfectly straight when the quill is flexed. When the feather is wrapped around the hook, the barbs will fully extend. This phenomenon needs to be taken into consideration when judging size. As a rule of thumb, a high-quality saddle feather will come out one hook size larger than it appears when merely flexed for sizing.

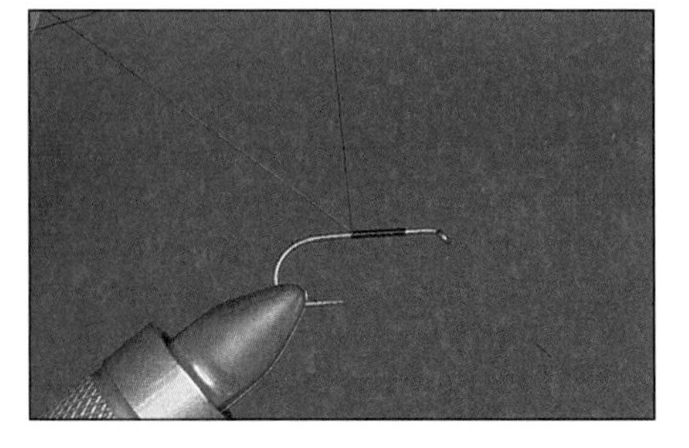

***Tying on; step 1.***

***Sizing hackle; step 2.***

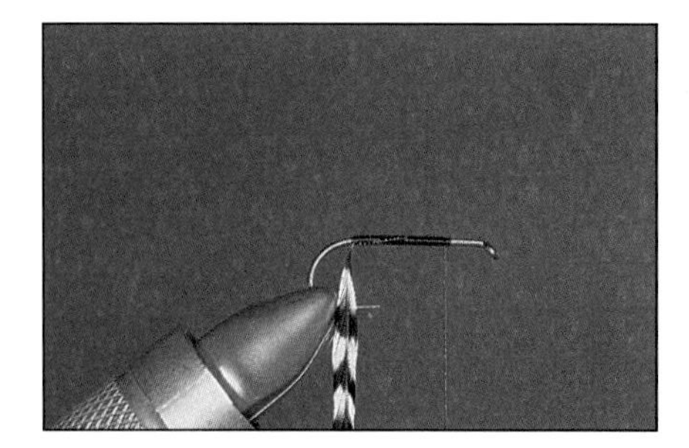

***Hackle tied in; steps 3-4.***

3. Take hold of the feather near the butt and pull it off the pelt. Strip material off the butt end until you encounter quality hackle. With top-quality saddles, dry-fly hackle may start almost at the butt end and continue throughout. If soft stuff runs well up the feather, though, simply cut back nearly to where the good material begins and start stripping from there. Do one side at a time to avoid breaking the quill. Don't "flare" the hackles, as is commonly done for wet flies; this is not good for the health of dry-fly hackle.
4. Tie in the feather *by its butt end,* beneath the hook, at the rear, *pretty-side-forward. Important: Leave a tiny bit of bare quill exposed.* This allows the quill to rotate into a 90-degree attitude to the shank before barbs begin peeling off, which guards against having the first half-turn lie at a rearward angle. You want the hackle

to stand out at just about a right angle to the shank. Then wrap the thread forward with neat, contiguous wraps, binding down the quill as you go. When you've reached a point about 1/4 shank length rearward of the eye, stop and trim off the tag end of the quill.

5. Pick up the feather and start wrapping. If you're using one of those long, good-quality saddle hackles, you won't need hackle pliers. Each turn should lie contiguous to the one before. Be sure to keep the pretty side of the feather facing forward; this is very important. If you're wrapping by hand and the feather becomes too short to grip comfortably, attach your hackle pliers and continue. Ain't dry-fly hackling fun?

***Wrapping the hackle; step 5.***

***Tie-off point for grizzly hackle; step 6.***

***Grizzly hackle butt trimmed off, white hackle tied in; steps 7-8.***

6. When you reach the thread, tie off the feather. If you have good, stout hackle pliers, you may wish to use them now—whether or not you used them in the wrapping—as they'll help you tie off hackle feathers. Clamp them onto the feather as shown, and let their weight maintain tension on the quill. Work the thread in among the barbs and take four or five very firm thread wraps behind the feather to bind the quill to the hook shank. Try not to tie down any barbs if you can help it.
7. Trim off the rest of the feather by cutting the quill. If there's enough feather left for additional flies, do this carefully. With today's saddles, it's not uncommon to get several flies from a single feather.

***White hackle in place; the completed Grizzly Bivisible.***

8. Select a white feather with the same length of barb as the main hackle feather's. It can be either a cape or saddle feather. Prepare it exactly as you did the grizzly, and tie it in beneath the hook shank in such a position that the first turn will abut the last turn of the grizzly. Remember: pretty-side-forward, and keep a tiny length of quill exposed.
9. Advance the thread to the eye, then back two or three turns. Wrap the hackle forward as you've been doing and, when you reach the thread, tie it off as you did the first hackle, and trim the excess. Secure it with a few firm housekeeping wraps, whip-finish, and lacquer.

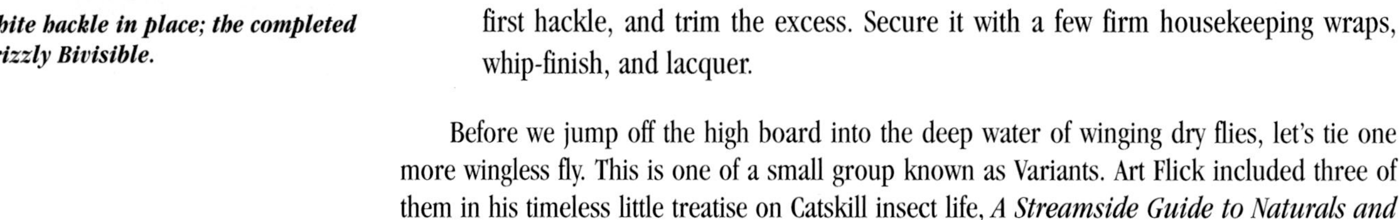

Before we jump off the high board into the deep water of winging dry flies, let's tie one more wingless fly. This is one of a small group known as Variants. Art Flick included three of them in his timeless little treatise on Catskill insect life, *A Streamside Guide to Naturals and Their Imitations:* the Dun Variant, the Gray Fox Variant, and the Cream Variant. We'll tie the last one here.

## Tying Terms

***Variant*** This term has several meanings. Here, it identifies a distinct design of dry fly that is characterized by an oversized hackle and a lack of wings.

***Stripped Quill*** This is a hackle quill from which the barbs have been removed. The term can also refer to a peacock quill that's been stripped of its herl. At one time, prestripped quills could be bought in strung bundles; they may still be sold that way, but I haven't been able to find them for a while.

As preparation, I recommend that the quills be stripped ahead of time and placed in a little bottle, such as an empty spice jar, in a mixture of 25 percent hair conditioner and 75 percent water. This will soften them and keep them from breaking while they're being wrapped. They can be left in the solution indefinitely—a year, at least.

***Zap-A-Gap*** This is a type of superglue that comes in very handy in fly tying. There are other brands, but as of this writing Zap-A-Gap is still my choice. It's somewhat less runny and uncontrollable than many of the other cyanoacrylates, and it is impervious to water.

One further note: With flies of this type, you'll be wrapping very large hackle feathers taken from well back into a cape. Therefore, you can expect aberrations when wrapping, such as barbs leaning forward or flaring off at various angles. Don't let this bother you; it's not your fault. Simply do the best you can, and neaten up with some discrete trimming later.

When Art Flick and his contemporaries were tying these monstrous flies many years ago, the hackle scene was entirely different. Few capes tied smaller than size 12 or 14 without trimming, and larger feathers were the rule rather than the exception. A half-century of selective breeding has turned that situation completely around.

If you happen to become a Variant addict, you might consider buying some cheap imported capes. These are available from certain fly shops and catalogs at a remarkably low price—which is about what they're worth. However, some have long-barbed hackles of reasonably good quality. Be sure to inspect them carefully, and don't hesitate to exercise your right of return if a catalog purchase turns out to be a disappointment.

***The Dressing*** Cream Variant

*Hook:* Daiichi #1180 or comparable.

*Hook Size:* For this exercise, size 14 or 16.

*Thread:* 8/0 Uni-Thread or comparable; light brown, tan, or yellow.

*Tail:* Long hackle fibers; straw cream.

*Body:* Large stripped hackle quill; light ginger or straw cream.

*Hackle:* Two or three hook sizes oversized; straw cream.

**Tying Steps**

1. Tie on at about the midpoint of the hook and wrap to the rear.
2. From the edge of a cream cape, select a spade hackle that has long, stiff barbs. Strip the softer material from the lower part of the quill, working up until you encounter the stiff barbs. Check to see if the barbs are of equal length on both sides of the

quill. If they are, barb-gather—as you've already learned to do—securing a bunch of about a dozen fibers for the tail. If the barbs differ in length on each side of the quill, work with one side at a time to ensure uniform barb length.

3. The tail on a Variant is longer than normal for its hook size, because it must balance with such long hackle. Gauge it at about $1\frac{1}{2}$ times the shank length, and tie it in with several pinch-wraps. Don't cut off the butts yet.

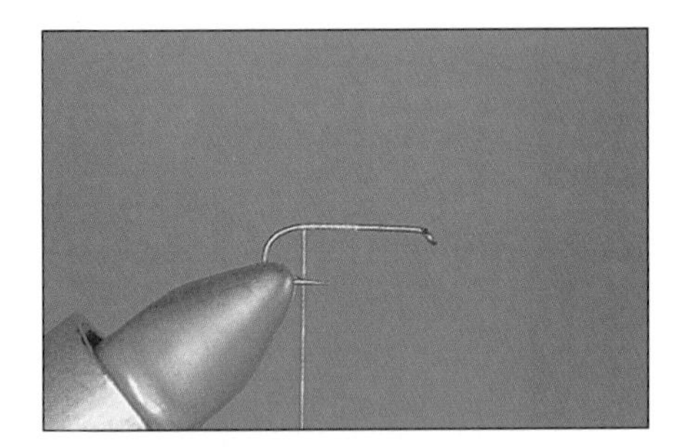

***The starting point; step 1.***

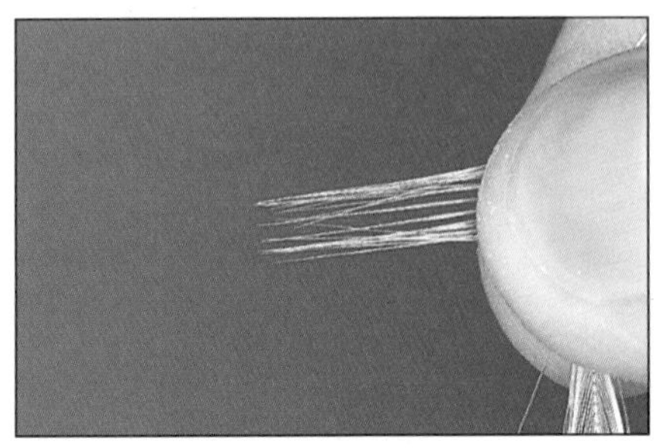

***Gathering barbs for tail; step 2.***

***Tail in place; step 3.***

***Binding down tailing and quill butts; step 4.***

4. If you've prepared the stripped quills per the instructions above, remove the softening solution from one by wiping it dry on a paper towel. Tie it in by its skinny end; then bind down both the tag end of the quill and the tail butts, working forward with neat, contiguous wraps. When you reach midshank, trim them both on a slope, and take a few more wraps forward. When you are $\frac{1}{3}$ shank length from the eye, stop.
5. The proportions of a Variant are about $\frac{2}{3}$ body to $\frac{1}{3}$ hackle. Wrap the quill with contiguous turns to the point where you left the thread waiting. Tie it down, trim the tag end on a slope, and secure it with a few firm wraps.

***Wrapping the quill body; step 5.***

6. Pick out two oversized hackles from the same cape that gave you your tailing material, or a cape of the same color. You'll see that the feathers around the edges have longer barbs and less web than those in the center. Strip back to the sweet spot, as previously defined. With large cape hackles, the high-quality portion will be approximately the outer $\frac{1}{4}$ of the feather. Don't worry, this is enough. And please don't try to cheat by using more of the feather than is prime. The heavier quill and poor-quality barbs will take up the space allocated for the prime stuff, and you'll hate the result.
7. Tie in these feathers at the front of the body, by their butt ends, leaving just a tiny length of quill exposed, as you did on the Bivisible. Work forward with firm, contiguous wraps. Before you reach the eye, trim off the butts on a slope; then wrap to the eye and back two or three turns.

***Two hackles tied in; steps 6-7.***

8. The hackles are wrapped one at a time, each wrap abutting the one before. When you wrap the first hackle, don't leave little spaces for the second one; it will find its own niches between the groves. When you reach the spot where you left the thread hanging, tie off the quill securely, trim the butt, and take a few housekeeping wraps. While wrapping the second hackle, if you find that it's mashing down any of the barbs from the first, back off and rewrap, wiggling the feather back and forth a little to get it to seat properly.

9. After tying off and trimming the butt of hackle number two, take a few wraps to secure everything, and finish off the head. Then take a toothpick and spread a thin layer of Zap-A-Gap over the quill body, covering all sides. Set it aside to dry.

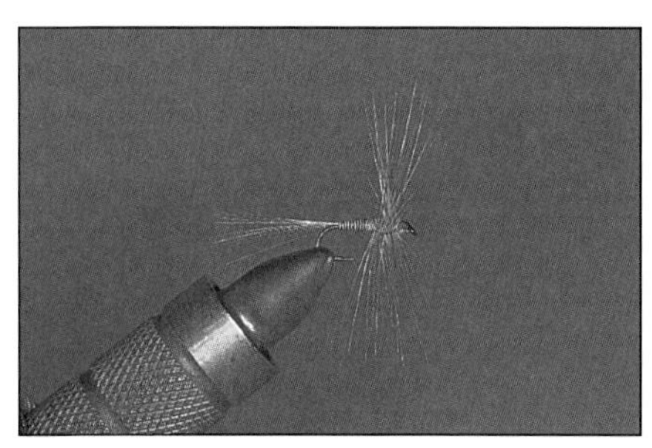

***Wrapping the hackles completes the Cream Variant; steps 8-9.***

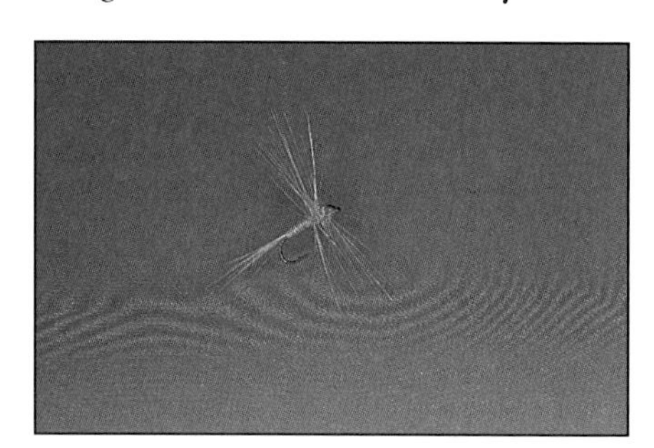

***Another view of the finished fly.***

The three Variants in Art Flick's book were tied to imitate specific mayflies that appeared on Art's home waters. The Gray Fox Variant represented the eastern green drake; the Cream Variant, the large, pale mayfly known as the *Potomanthus;* and the Dun Variant, the Leadwing Coachman, of which there are several subspecies. Like Bivisibles, Variants are great riffle and pocket-water flies.

Now I think it's time to begin learning the various types of dry-fly wings. There are two general categories: *upright wings,* which are the most common, and *down-wings,* which lie over the top of the body, almost streamer style. Within each category is a number of types of wings, each of which calls for a different material or technique. Here you'll tie the most commonly used wings; once you can tie these, the rest will be well within your capabilities. First, the one I consider the easiest.

**Tying Term**

***Tippet Wings*** As the name suggests, these are formed out of small hackle feathers called *tippets.* Lots of web and opacity are desirable attributes, and so tippets from hen capes are usually used. You can also use hen saddles, but you must be careful, as some hen saddle feathers have skewed quills that prevent the wings from going on straight. You may also encounter this problem in hen cape tippets, but it's much less common.

Tippets are used both in their natural form—as typified by the Adams wing—and in modified, or shaped, form. Shaping is done either by trimming or by using a wing burner, a tool I described in the first chapter. A good, sharp toenail clipper does a fair job of shaping tippet wings, because its curvature is constant. Curved scissors are okay but rather tedious to use; it's also difficult to make the wings uniform with them.

Wing burners, which are depicted in chapter 1, do a nice job. All that's required is a butane lighter and feathers with broad, webby centers. Simply mount each feather in the tool with the quill centered, and burn away everything that sticks out around the edges.

I suggest you look for a set of wing burners that's as close to symmetrical as possible. This makes it much easier for you to produce pairs of wings that are virtually identical. In fact, symmetrical wing burners allow you to make both wings at once. Provided the quills are centered in the tool, the wings will come out as mirror images. To keep matters simple, we'll use tippets in their natural form for the following fly.

Other than the wings, this fly presents no materials or techniques that you haven't already encountered in this book.

***The Dressing*** Blue-Winged Olive

*Hook:* Daiichi #1180 or comparable.

*Hook Size:* For this exercise, size 14.

*Thread:* 8/0 Uni-Thread or comparable; olive.

*Wings:* Gray hen hackle tippets.

*Tail:* Gray barbs from a cock spade feather.

*Body:* Very fine-packing dubbing; olive.

*Hackle:* Dry-fly grade; gray.

## Tying Steps

1. The wings go on first. To prepare the hook surface for them, tie on a short way behind the eye and wrap thread contiguously rearward to almost the midpoint of the hook, then forward to the tie-on point, then rearward again to the center of the thread base you've just established. This positions the thread about 30 percent of the shank length to the rear of the eye.

2. Select two tippets from a hen cape that are as close to identical as possible. When you're estimating size, keep in mind that after you've stripped them of the waste material at the base, the tippets need to be long enough to make wings that are about the length of the hook shank. Then manicure the feathers so that they resemble those in the photo. Leave them a bit long, for reasons I'm about to explain.

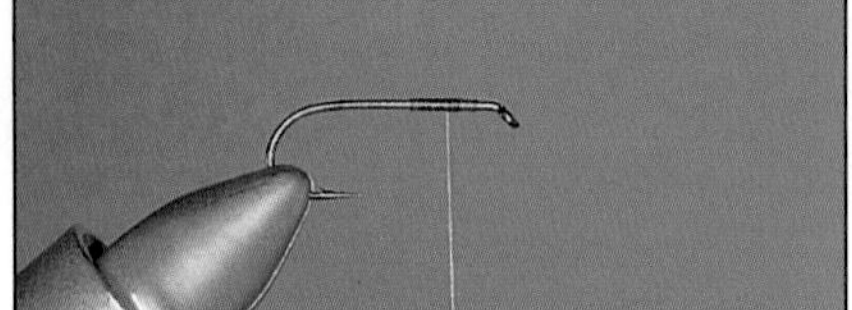

***Thread base for wings; step 1.***

***Hen cape tippets; step 2.***

3. Hold the tippets back-to-back, so that their natural curvature causes them to flare away from each other. With your left thumb and forefinger, hold them over the hook, and gauge the length. Then take them by the tips with your right thumb and forefinger, and carefully stroke back—but don't strip off—the barbs, until the desired length is established. Keep the tips even throughout, so that both wings come out the same length.

4. With your left hand, hold the tippets as shown, isolating the parts that will form the wings from the parts that you stroked back. Position the tippets on the thread base and tie them in place with several pinch-wraps. The thread intersects the quills precisely where they are exposed.

5. Take a few more firm wraps immediately behind the wings, and inspect them. If they've slipped toward the far side, simply move them back on top and recenter them. This is easy to do, as the wraps you've used to this point have not completely

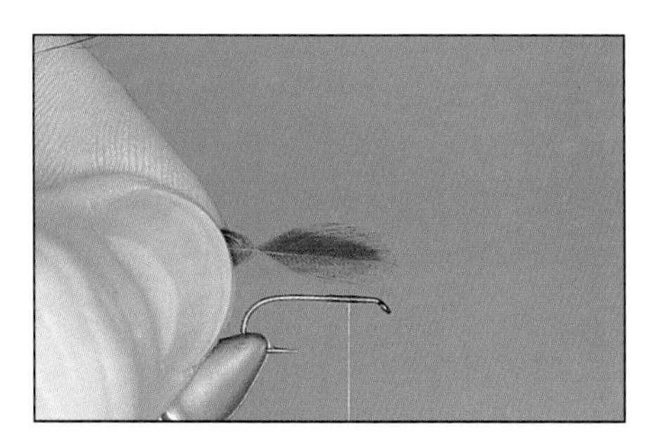

***Hen tippets prepared for tie-in; step 3.***

***Tying on tippets with the pinch method; step 4.***

locked in the feathers. Just stroke them upward, simulating the attitude they'll assume in their final position.

6. When the wings are properly positioned, lock them in with several more tight thread wraps immediately to the rear. Trim off the butts with a slope-cut.
7. Stand the wings up and, with your thumb, gently crimp the quills at the base. Then bring the thread forward under the hook and take several turns against the butts, forcing the wings into an upright position. Use only as many turns as are required; you don't want a lot of clutter in this area.
8. Wrap the thread back underneath the hook and to the bend, binding down the trimmed butts in the process.

***Wings upright and manicured; steps 5-8.***

9. The tail is tied in the same manner as that of the Variant, except that it isn't quite as long. It should be the length of the hook shank, or perhaps just a tiny bit more. On a flat surface, a conventional dry fly should balance on the tips of the hackle and tail, with the hook either just touching or barely clearing the surface. Try to integrate the butts of the tailing barbs with the butts of the wing material, so as to create a smooth underbody. Bind them down with very neat thread wraps, working forward to the point where the wing butts end, then rearward again, stopping a few turns shy of the bend.

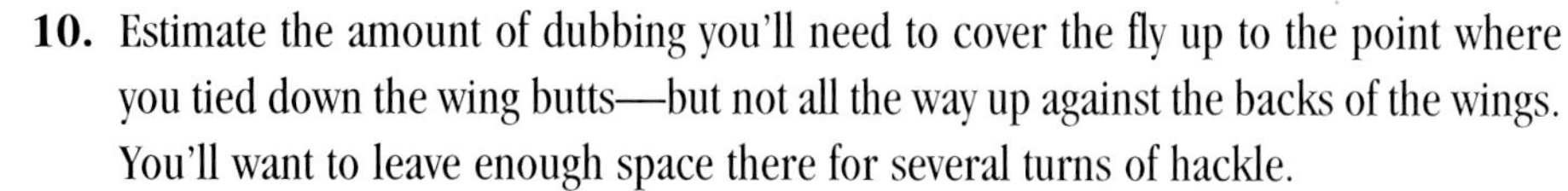

10. Estimate the amount of dubbing you'll need to cover the fly up to the point where you tied down the wing butts—but not all the way up against the backs of the wings. You'll want to leave enough space there for several turns of hackle.
11. Spin a thin worm of dubbing onto the thread. If you can taper it in such a manner that it gradually builds toward the front, so much the better. Try to work the dubbing up quite close to the hook shank, so that when you begin to wrap, you can take a few turns to the rear and the dubbing will begin to deploy just as the bend is reached. If you have difficulty working this close to the hook, try placing the fingertips of your right hand under the dubbing, after you've finished spinning it onto the thread, and sliding it up the thread toward the hook.
12. Wrap the dubbing, leaving space to the rear of the wings, as stated.
13. Select two hackles, gauging for standard dry-fly proportions: The barb length should equal 1½ hook gapes. Prepare them by stripping back to their sweet spots. On a typical hackle feather of good-to-prime quality in this size range, the sweet

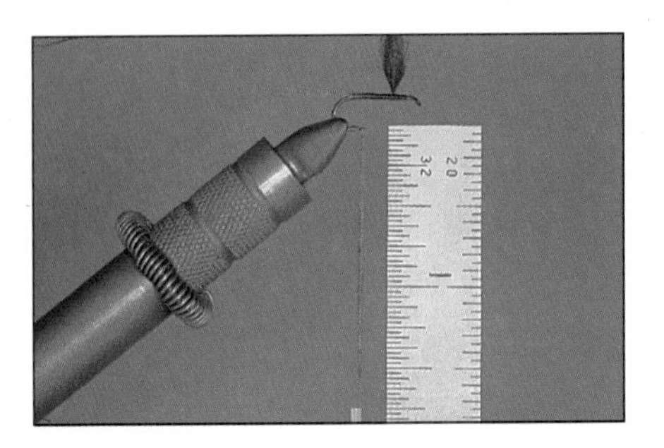

***Tail in place, amount of dubbing being determined; steps 9-11.***

***Body in place; note space rearward of wings; step 12.***

spot will be $1\frac{1}{4}$ to $1\frac{1}{2}$ inches (30–37mm) in length, and you can expect to get from five to nine turns of hackle. (This number will vary a bit due to the factors I discussed in "Judging Quality," "Grading," and "Capes or Saddles?"). Tie in the feathers beneath the hook, just to the rear of the wings, pretty-side-forward and with a tiny bit of quill exposed.

14. You'll probably notice that one of the feathers is positioned in such a manner that it can be wrapped first, without its quill crossing over that of the other feather. Wrap this hackle first. Leave just enough space for one turn of the second hackle at the very front of the body; but other than that, make the wraps contiguous.

15. You'll be able to take from two to four wraps behind the wing, three being just about ideal. Don't try to sneak in an extra turn here, as you'll crowd the wings and cause them to lean forward. Come directly beneath the wings, and take the first turn of hackle in front, tight to the wing base; this helps support them in an upright position. Wrap forward, using up the rest of the sweet spot. You should end up a few thread wraps behind the eye, which will leave adequate space for your tie-offs and whip-finish.

***First hackle being wrapped behind wings; steps 13-15.***

16. Tie off the first hackle, trim, and secure with a few housekeeping wraps. Then pick up the second hackle, and wrap it right through the first one. As I mentioned in the Variant exercise, if any matting occurs, back off and reposition.

***First hackle wrapped and tied off; steps 15-16.***

***Second hackle in place, completing the Blue-Winged Olive; steps 16-17.***

***Another view of the finished fly.***

17. Tie off the second hackle, trim, and clean up the area with a little discrete scissor work, blocking off the thread with your left forefinger. Whip-finish, and you've completed your first winged dry fly. Gorgeous, isn't it? I'm sure the fish will agree.

The next exercise will teach you the famous and time-tested Wulff-style hair wing. This wing utilizes the same technique as the equally famous wood duck wing. For years, I taught beginners the wood duck version first, because I felt it was a little easier to handle that material than hair. Having implemented a couple of innovations in my own methodology, though, I now teach the hair version first.

### *About Hairs for Dry-Fly Wings*

Circa 1930, Lee Wulff created the fly design that became his namesake. I'm not sure if the original wings were made of calf tail or bucktail, but I do know that, many years later, when I became acquainted with him, Lee was using calf tail. He liked his wings bushy and somewhat unkempt, and he sometimes chided me gently for making mine too neat. I took this as a compliment, albeit a left-handed one.

The hairs most commonly used in tying this style of wing today are, in no special order, calf tail, calf body hair, and deer tail or bucktail. I've also seen a few other types of hair used for certain patterns: moose body or woodchuck tail are sometimes used for the Black Wulff, with Norwegian goat substituting for the calf and deer hairs.

Calf tail has crinkly hair, which produces a bushy sort of wing. Calf body hair is finer in texture and much straighter. It's also shorter, which limits the size of fly for which it can be used. However, prime calf body hair from a winter pelt can easily accommodate quite a sizable Wulff, and it's perhaps the most user-friendly hair of all.

Deer tail is interesting. At first glance, it's difficult to picture it as being adaptable to this sort of application. However, when it's properly manicured and run through a hair evener (stacker), it lends itself quite nicely to this task. The resulting wing will be much neater in appearance than one made of calf tail, which means that you, the tyer, will have to decide what the character of your Wulffs is to be.

Some tyers curse the hair wing for its perversity, but actually, it's not at all difficult once you understand a few things. The main factors in successful Wulff-winging are material selection, material preparation, and tying technique. I might add that if the first two aren't attended to, the latter will be sorely tested.

When buying calf tail, look for fairly straight hairs; they're all kinky to a degree, but some are kinkier than others. Great length isn't that important; the hairs need be only long enough to be manageable during the winging process. Leave the exceptionally long haired tails for the streamer tyers.

It's important to clean all short and aberrant hairs from a winging bunch. You can do this with your fingers or with the assistance of a little comb, which I heartily recommend. When the bunch of hair is ready to be tied in place, nothing should remain that doesn't contribute to the wing silhouette.

The next advisory applies to all hairs involved in Wulff-winging: Don't try to use too much material. This is a very common mistake and one that causes a lot of pain. Use just enough to attain an attractive, visible wing, and no more. Too much hair not only causes terrible tying problems, it also results in a clumsy, poorly balanced fly.

Stacking, meaning using a hair evener, is also extremely important. Not only does it optimally position all hairs, it also enables you to accurately judge quantities. You really don't know how much material you have until you've evened up the hairs. Years ago, I read an article in one of the major fly-fishing publications claiming that calf tail couldn't be stacked. Not so! I'll show you how in the exercise.

The only other component that differs appreciably in this fly from what you've done to up to now is the body. This one is unique, and it's simply known as the Royal Coachman–style body. It's not hard—just different. Here, we'll tie a variation of the Royal Coachman that I was introduced to in New Zealand in the fall of 1993. It didn't seem to have a name, so I've decided to call it the SteppenWulff, after that great literary classic.

## Tying Terms

***X-Wraps*** These are used to divide the wings. The process will be described in the tying steps.

***Figure-8 Wraps*** These are used to form the wings and lock them into position. They will also be described as we go.

**The Dressing:** SteppenWulff

*Hook:* Daiichi #1180 or comparable.

*Hook Size:* For this exercise, size 10 or 12.

*Thread:* 8/0 Uni-Thread or comparable; black.

*Wings:* White calf tail.

*Tail:* Brown hackle, brown bucktail, or woodchuck tail hair.

*Body:* In three segments: peacock herl; a center band of bright green floss or floss substitute; herl again.

*Hackle:* Mixed brown and grizzly.

*Necessary Tool:* Wide-tube stacker.

### Tying Steps

1. Tie on and create a thread base for the wings just as was done for the tippet-wing fly.
2. Hold a bunch of calf tail hairs at a 90-degree angle to the tail itself, and cut them off at the base.
3. Begin the cleaning and manicuring process by stroking or combing out all the underfur and short hairs. Hold onto the tips, and work toward the butt ends.
4. When you've weeded out all of the junk, reverse your grip and, while holding the hairs by their butt ends, pass your little comb through the tip ends a few times. You can't straighten the kinky fibers, but you can unlock them.

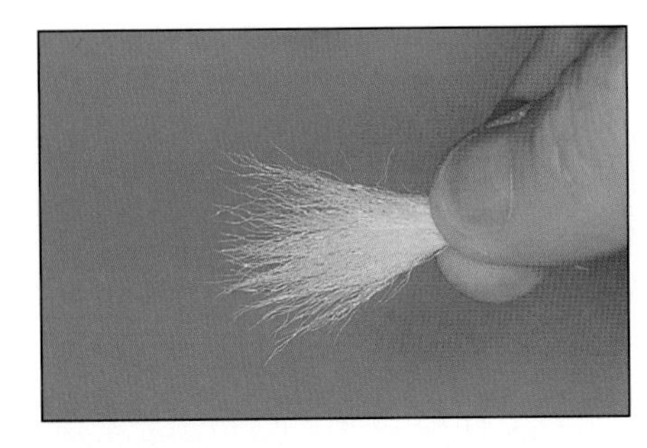

***A bunch of calf tail hair for Wulff-winging; step 2.***

***Cleaning the bunch; step 3.***

***Tips evened. Reverse the direction of the stacker tube before removing the hair bunch; step 5.***

5. Insert the hairs tips-down into a wide-tube stacker. Knock it on the table 10 or 12 times. Then hold the tool horizontal and face it so that when the tube is slid out, the tips of the hairs will be pointing forward.
6. Remove the stacked bunch from the tube, pass it to your left thumb and forefinger, and measure the wing length against the hook shank. With a Wulff wing, it's doubly important to adhere to proportions: This material is heavy, as dry-fly materials go, and too-high wings will unbalance the fly.

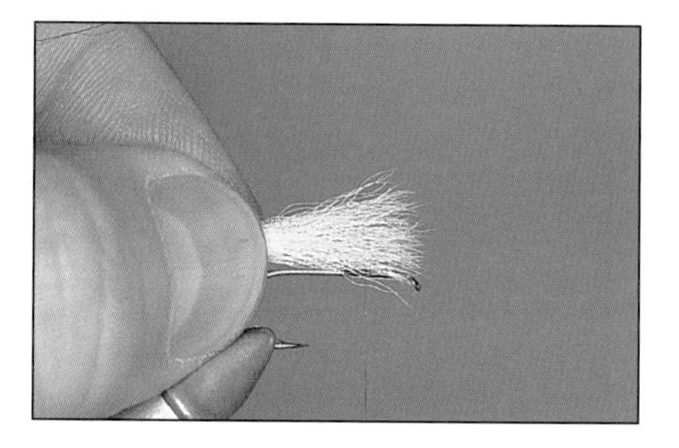

***Thread base in place, wing length being established; step 6.***

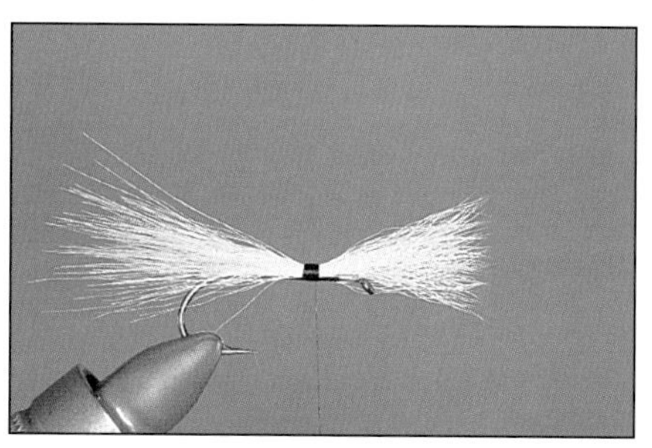

***Winging bunch tied in place; step 7.***

***Hair butts trimmed and bound down; step 8.***

7. Tie the wing onto the thread base using a series of pinch-wraps and very firm securing wraps. Before the winging bunch is absolutely immobilized, stroke it temporarily into an upright position, bringing any hairs that have rolled to the rear back on top, where they belong.
8. When you're satisfied that all the hairs are centered, secure them in place with a few more tight wraps. Then, before you stand the bunch up, trim the butts on an elongated slope. They will later integrate with the tail butts to balance out the underbody. Wrap back and then forward to cover the trimmed ends, finishing with the thread at the original tie-in point of the bunch of hair.

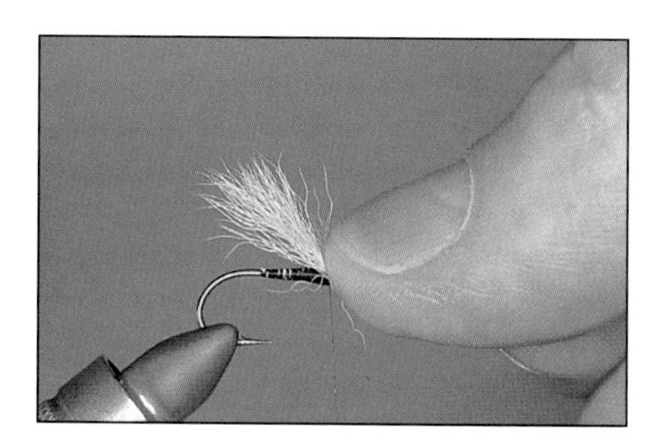

***Crimping the hair; step 9.***

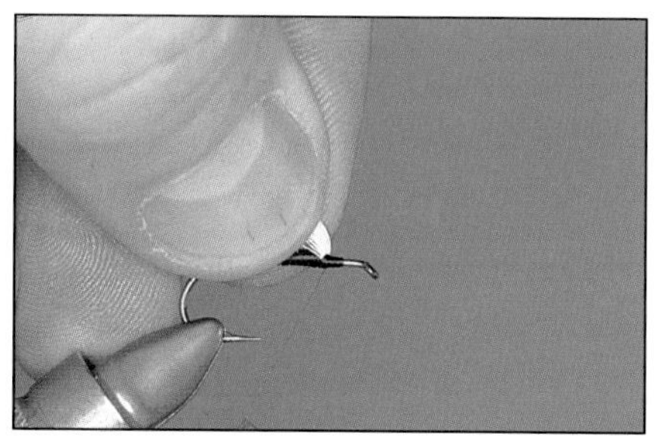

***Thread "dam" stands up the hair; step 9.***

9. Lift the hair upright and crimp it a few times with your right thumb to give it a bit of "memory." Then hold it upright with your left hand and bang lots of thread wraps tight to the base, building a virtual thread "dam." Work back and forth a little, so that the thread doesn't all lump up in one spot. When the hair stands perfectly upright by itself, stop.
10. Using any pointed instrument, divide the hair into two equal bunches. It helps to

look down from above while doing this. If your vise has revolving capability, it will come in handy now. Separate the two bunches by pulling them firmly into a spread position.

11. To divide the wings, take two X-wraps. Start with the thread just in front of the wings, pass it through the V between the wings, then wrap around and under the hook, just behind the wings. Now bring the thread up behind the near wing and pass it forward through the V, then down. This places an X of thread between the wings and positions the thread back where the procedure began. Repeat it.

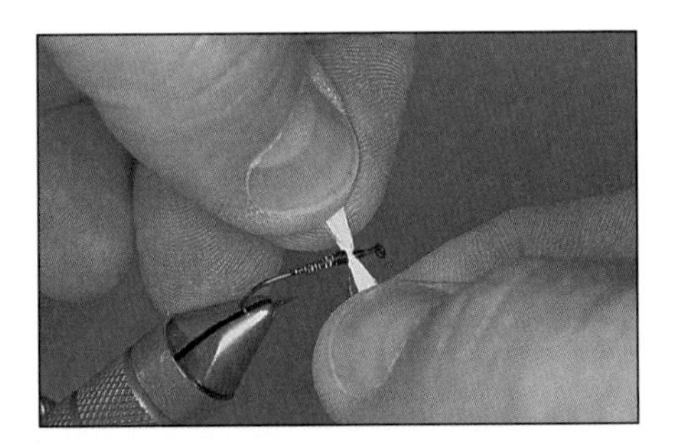

***Dividing the hair; step 10.***

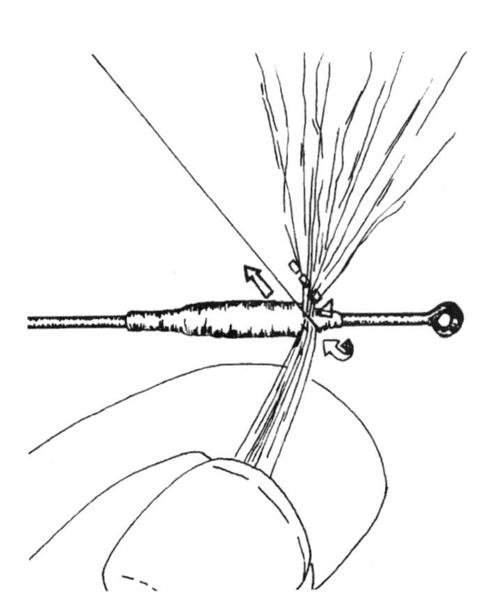

***The beginning of the X-wrap.***

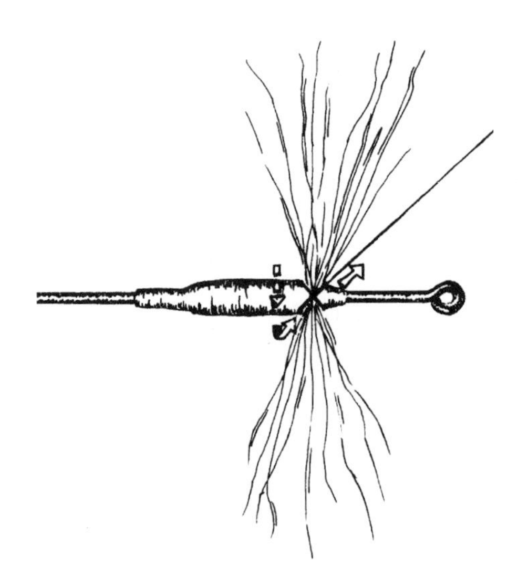

***The completion of the X-wrap.***

***The X-wrap; step 11.***

**12.** The second part of the procedure is called the figure-8. Begin it as you did the X-wrap, but when the thread is just below horizontal behind the far wing, seize the wing with your left thumb and forefinger. Hold it tight and pass the thread, under firm tension, all the way around the base of the wing. Then bring the thread back through the V between the wings and down behind the near wing. Throughout this entire sequence, be sure to maintain thread tension and to grip the wing firmly until the figure-8 is finished, with the thread hanging straight down. Study the illustrations closely.

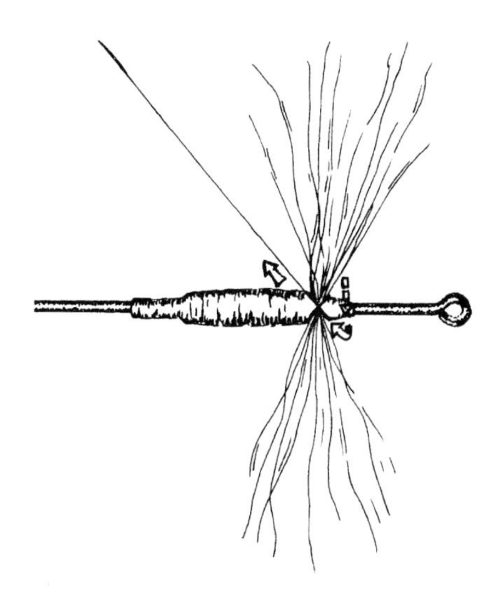

***To begin the figure-8, come around under the hook and between the wings, as though making another X-wrap.***

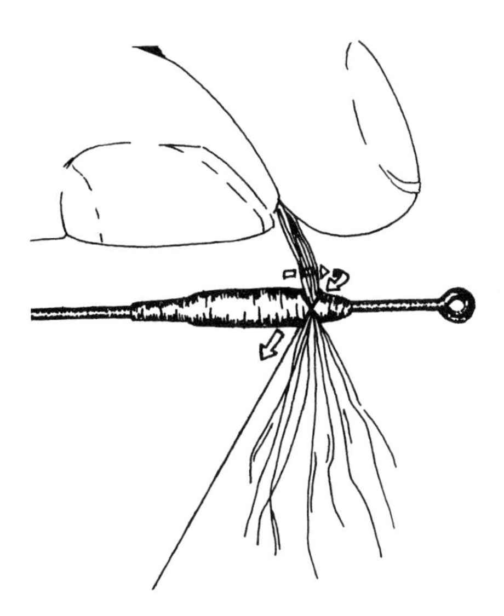

***Come around the base of the far wing and back through the notch.***

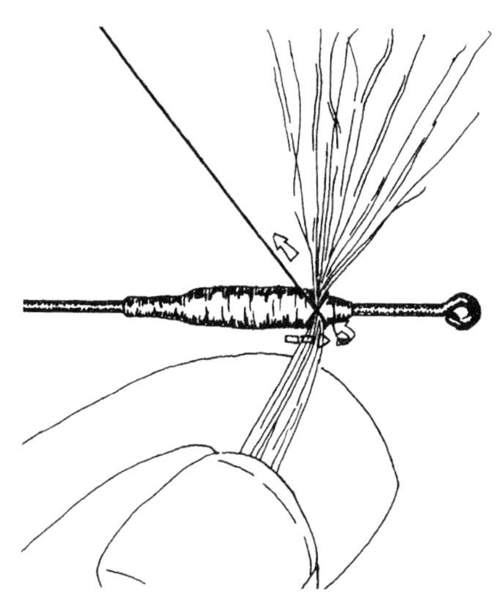

***Come around the base of the near wing and back through the notch.***

**13.** Now seize the near wing. Wrap the thread all the way around the base of the wing, back through the V, and down behind the far wing. This completes the figure-8. Put another one on top of it, and—in the case of very large Wulffs with heavier wings—a third.

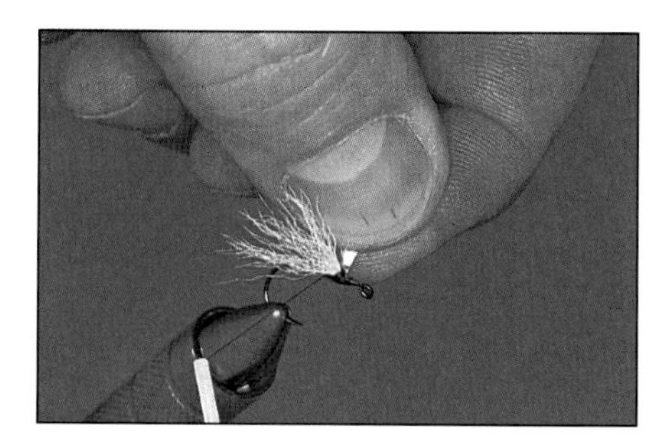

***Figure-8 starts around base of far wing; step 12.***

***Figure-8 concludes around near wing; step 13.***

***Appearance of completed wings; step 13.***

**14.** Wrap back to the bend, and procure a bunch of tailing material. Of the ones discussed above, I suggest hair for larger Wulffs and hackle for those size 14 and smaller. Here, we'll use hair. Stack it as you did the wings, tie it in just as you've been doing, and use the same proportions. For a hackle tail, proceed as with previous flies.

**15.** Trim the tail butts on a slope, so that they complement the wing butts. Bind them down with neat wraps, working forward, then back to the bend. Do not put any more thread around the spot where the hackle will be wrapped.

**16.** Before proceeding, you may wish to review the section on peacock herl in chapter 4. Select four or five fronds, cut them back a bit at the tips, and tie them in by the tip ends.

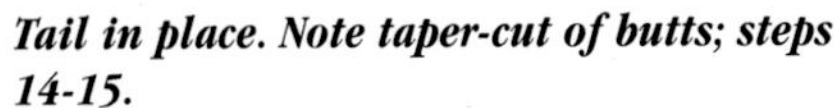

***Tail in place. Note taper-cut of butts; steps 14-15.***

***Peacock herl tied in; step 16.***

***Rear body segment completed; steps 16-17.***

**17.** Make a large loop with the thread, cut off one side to reduce it to one strand of thread, and twist it in with the herl. Take just two or three turns, forming a bump, as shown. Then bind the herl and thread to the shank for a short way, creating a base for the center band. Leave plenty of space for the front peacock bump and for hackle behind the wing.

**18.** Cut off a short piece of the floss and tie it in to the front of where the center band will be located. Wrap it rearward a couple of turns, then forward over itself, and tie it off. When trimming the tag end, leave a little extra, so that the floss can be further secured as you wrap the thread forward to its next position.

**19.** Pick up the herl again, twist to mix it with the thread, and make the front bump with a couple of wraps. Tie it off well clear of the rear of the wings, and trim off the tag ends. *Option:* If you wish, you can roll the herl butts up onto the top of the hook, in front of the wing, and bind them down almost to the eye before trimming. This helps compensate for the disparity in plane caused by the tail butts. It's a small refinement that will make your hackle come out a bit more even.

***Floss "belly band" in place; step 18.***

***Front body segment completed; step 19.***

**20.** All that's left for you to do now is apply the hackle. Do this just as you did for the BWO. You can use either cape or saddle hackles, whichever you happen to have. You can even use one of each if they match up well. The important thing is that the length of their barbs is equal. When mixing brown and grizzly, I generally tie in with

*Hackles tied in place; step 20.*

*Brown hackle wrapped; step 20.*

*Two views of the completed SteppenWulff.*

the brown in front and wrap it first, because brown hackle is usually a little thicker of quill than the grizzly. If you find this isn't the case with the feathers you happen to be using, reverse the order. When you've completed the hackle, trim, whip-finish, and you have yourself a magnificent Wulff.

Now that you can tie the Wulff wing, you should be able to tie the wood duck wing with little difficulty. The fly in the following exercise is a sort of mix-and-match pattern that I use in my classes, because it teaches several important lessons at once. It's also a very good general-purpose fly astream.

### *About Waterfowl Flank Feathers*

In the wet-fly section, you used duck flank feathers to tie wet-fly wings. Now you'll see how they adapt to the dry fly. The most beautiful and coveted of all is drake wood duck. These are the delicately barred, lemon-gray feathers that lie under and behind the bird's wings. Only the male of the species has them.

There are two kinds of flank feathers on each side of the bird; the plain lemon-barred kind and the black-and-white-tipped kind. The latter are widely used in salmon-fly tying and on certain streamers.

Wood duck is generally in short supply and rather pricey. It shouldn't be, as there are plenty of wood ducks around, and many are killed by hunters during the season. Unfortunately, very few of these make it to the tying table. Like all waterfowl, wood ducks are federally regulated, and the process you must endure to obtain the license necessary to sell them legally is quite burdensome. Thus, most hunters, guides, and outfitters simply throw the feathers away. Horrors!

If you hunt in wood duck territory, or have friends who do, here's an easy way to take care of the feathers. Have four Zip-loc bags ready. Simply pluck the flank feathers from one side of the bird at a time. Bag the plain feathers and tipped feathers separately. Do the same with the other side of the bird. Mark the bags "left" and "right," and staple or clip them together.

Actually, it wouldn't much matter if the plain flank feathers were all thrown into one bag. However, the tipped ones must be kept separate, because of the manner in which they'll be used. You may never tie salmon flies, but I can promise you that anyone who does will greatly appreciate such a gift. And incidentally, you might be able to swap these feathers for something quite desirable.

In addition to wood duck, teal and mallard flanks are commonly used for dry-fly wings. I prefer teal to mallard, as the shape and texture are a bit closer to those of wood duck. Teal feathers' natural coloration is white with dark gray or black barring. As we've seen, they are often dyed to the wood duck shade. Properly done, this produces an acceptable substitute. Both teal and mallard are also used in their natural state. In fact, you'll be using teal in the forthcoming exercise; I don't want to send you on a frustrating, costly, and possibly fruitless search for real wood duck just yet.

Both mallard and teal are readily available in fly shops in bags of various sizes. Buy plenty; they're cheap, and there's quite a lot of waste. However, if you tie various types of flies that use mallard or teal, you'll find that many of the feathers that aren't suitable for dry-fly work are fine for other applications.

### *About Flank-Feather Wing Construction*

There are some things you need to know about this type of wing. Basically, there are three ways to tie it: the *single-feather full-tip method,* the *opposing-sections method,* and the *two-feather method.* They all have their place. Choice of method is not based on preference alone; it also is dictated by the size and shape of the feathers at hand. In the exercise, we'll be using the single-feather method, which I consider the basic. Afterward, I'll briefly cover the others, which are merely variations.

### *About the Peacock Quill Body*

This fly employs a different type of quill body than did the Variant. The quill here is taken from the "eyed" portion of a peacock tail feather. The reason is that these quills, when stripped of their herl, have a light edge and a dark edge. When wrapped, they make truly beautiful bodies with realistic-looking segmentations. One the most famous dry flies of all time, the Quill Gordon, calls for such a body.

These quills do not give up their herl easily. Here are two methods for stripping them:

1. The eraser method. Cut off one individual quill and lay it on a piece of cardboard. With the more abrasive side of a pencil/ink eraser, rub the herl off each side.
2. The paraffin method. In a saucepan, melt a small chunk of paraffin with some water. It will lie on top of the surface of the water. Take the pan off the stove. Take a peacock eye and immerse it in the water. Slowly withdraw it over a period of a few seconds. This allows a thin coating of the paraffin to adhere to the feather. In a few minutes, after the paraffin has cooled, you can pick out individual quills and scrape off the herl with your thumbnail.

The latter is my method of preference. In either case, I suggest you prestrip as many quills as you'll need for the number of flies you want to tie at one sitting—plus a couple of spares, in case of breakage.

Looking at an eye from the front, you'll see that the quills to the left of the center have their dark edge on the opposite side of those to the right of center. This is of significance, because you must always tie in the quill in such a manner that the dark edge lies to the rear; this creates the segmentation. No problem: Simply tie on the ones from the left by their skinny ends and those from the right by the butt ends. Spacing is controlled by overlapping, or not overlapping, the wraps. Given a choice, I prefer the quills from the right side, because there's better contrast and more quill strength nearer the butt end.

Gorgeous as they are, peacock quill bodies tend to be delicate and require protection. This can easily be done by using a toothpick to coat the finished body with a thin layer of Zap-A-Gap. Let it dry completely, and be careful not to get any glue on the other components of the fly.

### *About the Tail*

The following applies to dry-fly tails in general and to this one in particular. As we've seen, the butts contribute to forming the underbody, and it's important that the underbody be as smooth as possible. This is true in spades with peacock-quill bodies; they hide nothing, and lumps and bumps in the underbody become just that in the finished one.

To counteract this, there are a couple of things you can do:

1. Try to find tailing fibers long enough that the butts extend forward to where the body will end. This length is ideal, because the butts can be bound down to form a smooth underbody.
2. Failing that, trim the wing butts a little farther back, so that the butts of the tailing barbs can be integrated with them. This is delicate work, so don't be discouraged if you find that some practice is required. Believe me, it'll be worth the effort.

Now that you've plowed through all that heavy reading, I'm sure you're ready to resume tying. The fly you're about to tie may have another name, although I couldn't find it in any of my references. If someone else has been tying it longer than I, and has given it a name, I apologize. I call it the Teal-Winged Adams Quill.

***The Dressing*** Teal-Winged Adams Quill

*Hook:* Daiichi #1180 or comparable.

*Hook Size:* For this exercise, size 12 or 14.

*Thread:* 8/0 Uni-Thread; brown or black.

*Wings:* Teal flank, tied wood duck style, the length of the hook shank.

*Tail:* Mixed hackle barbs; dark ginger or brown, and grizzly.

*Body:* Quill from a peacock eye, stripped.

*Hackle:* Mixed grizzly and dark ginger or brown.

## Tying Steps

1. Make the thread base per the instructions for preceding flies.
2. Select a teal flank feather with a centered quill and a nice, even, squared-off tip.
3. Strip the fibers from the sides until the feather resembles the one in the photo.
4. Inspect the feather. If the quill extends out near the tip, go back about ¾ inch and snip it off, removing as little of the feather as possible.

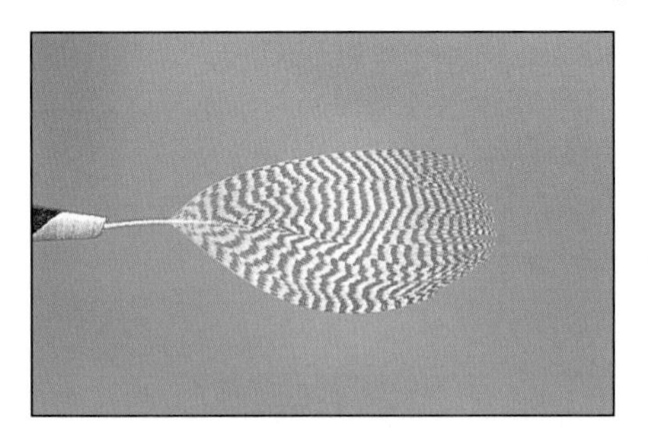

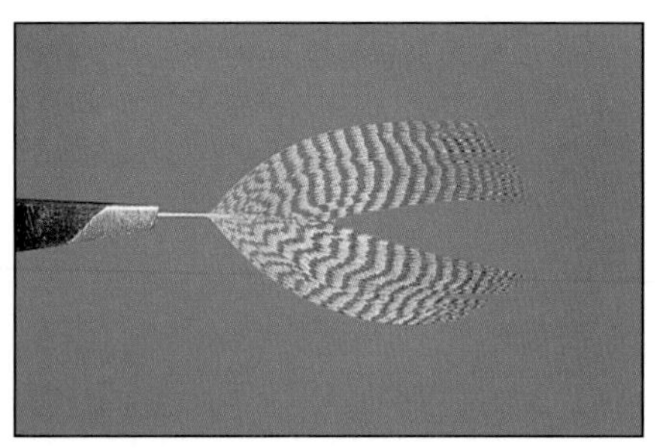

*Teal feather being prepared for use as wings; steps 2-4.*

*Gathering the teal fibers; step 5.*

5. Gather the feather in such a manner that it forms a neat bundle, with the tips even. This is done by stroking the fibers outward at an angle to the quill.
6. While holding the bundled feather with your left thumb and forefinger, gauge its length against the hook shank, then tie it onto the thread base with the usual pinches and securing wraps. Make corrections for torque, as I described in chapter 2.
7. Proceed just as for the Wulff wing: After securing the feather in position, trim the butts on a long slope, and either bind them down now or wait until you've completed the wing.

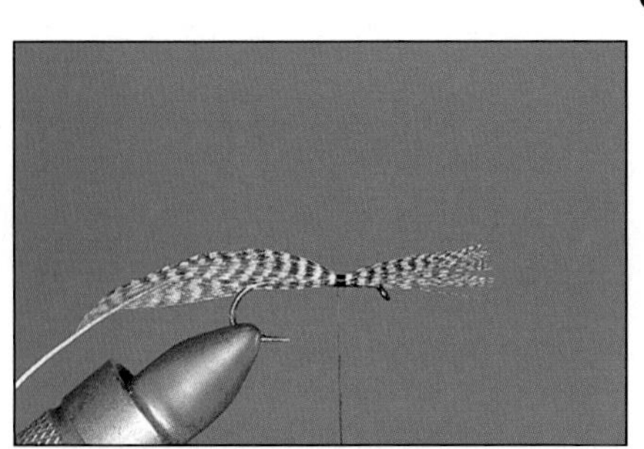

*Teal tied in; step 6.*

*Follow the same steps as for Wulff-winging to complete the flank-feather wings; step 7.*

8. From here on, the procedure is the same as for the Wulff: Stand the bunch up, separate it into two equal portions, take one X-wrap to divide, then two figure-8s to configure, and wrap very neatly to the bend.
9. This tail is no different than those you've been tying, except that it consists of mixed barbs. Here's an easy way to mix them. Do your barb-gathering on either the brown or grizzly feather; it doesn't matter which one you begin with. Cut off half of what you'll need for a tail, gauge for size, and tie on in the normal manner—but with only two turns of thread. Procure a second bunch of equal quantity from the other feather. Hold it in your right hand and lay it on top of the first bunch, getting all the tips even. Then grab both bunches with your left thumb and forefinger and unwrap the two turns of thread. Roll the two bunches between your fingers a little; then tie them on in the normal manner.

10. If possible, integrate the tail butts with the wing butts, trimming on a slope and using neat thread wraps. End up back at the bend.
11. Tie in the stripped peacock quill with the dark edge to the rear, so that when the quill is wrapped, it will be the trailing edge. Trim the tag end and advance the thread to where the body will finish. Pay attention to spacing; don't crowd the wing.

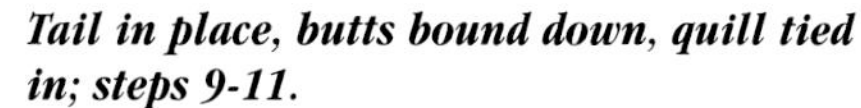

***Tail in place, butts bound down, quill tied in; steps 9-11.***

***The finished body, showing segmentation; step 12.***

12. Wrap the quill, creating light/dark segmentations, as shown. Unless you have an exceptional touch, you'll want to use hackle pliers. When you've completed the body, tie off and trim the remaining quill.
13. Finish off the fly by adding the hackle as usual; whip-finish and lacquer. I mentioned applying Zap-A-Gap to the body. If you haven't yet acquired that product, clear head cement will do.

***Hackling completes the Teal-Winged Adams Quill; two views.***

Wood duck is too valuable a commodity to waste. The professional tyers who worked during the glory days of the Catskills were extremely selective about their wood duck and very fussy about the technique for converting it into wings. To my recollection, both the Darbees (Harry and Elsie) and the Dettes (Walt and Winnie) adhered to the opposing-sections method. They usually used two feathers, because it's rare to find a wood duck flank with perfectly centered quill and optimal material on each side. Here's the process:

1. Cut two opposing sections from the prime sides of a pair of feathers. Reshape them a bit, so that the tips are squared off.
2. Hold them back-to-back, gauge length, and tie them in, taking care to maintain their relative positions.

3. Stand the fibers up, as you would any winging bunch. Then go in with a toothpick, or something, and separate the two sections, paying attention to maintaining opposing curvature.
4. Proceed to develop the wings in the normal manner.

The advantage of this method is delicacy: There's no quill involved, and using only the most prime sections optimizes silhouette while minimizing bulk. It does consume a lot of wood duck, though. I guess those tyers had access to a bountiful supply, but most of us don't, so we look for ways to use every last feather—even the little guys that end up in the bottom of the box. With apologies to my predecessors, here's how the two-feather method works:

1. Select two smallish wood duck feathers that aren't large enough to be made into a pair of wings individually. Strip off fibers until only the prime parts remain.
2. Hold them convex-to-convex, and proceed from this point just as though they were a pair of opposing sections: Compact them into a bunch, tie them on, and make the X- and figure-8 wraps. Try to maintain their relative positions throughout, because the opposing curvature results in a beautiful wing.

The potential problem here is that the two quills will contribute too much bulk to the wing butts—and if the wing butts are excessive, hackling is affected. To prevent this, avoid using feathers with heavier quills, and don't make your wings too bulky (a temptation when two feathers are involved).

If you can tie the flies presented up to this point with reasonable success, you're well on your way to becoming an expert. These are some of the most difficult processes in fly tying, short of married-wing salmon flies and such. Now, in order to complete what I consider to be the basic dry-fly skills list, you'll tie several more types. You'll be pleased to discover that they're considerably easier.

## The Comparadun Wing

Comparadun is a copyrighted name belonging to Al Caucci and Bob Nastasi, coauthors of *Hatches,* a fine book on stream entomology. It describes a type of fly attributed to them. There are several versions. Here, we'll tie the most common.

This is a simple fly to tie, if you can adhere to a couple of disciplines. The only two problems are avoiding too much bulk in the body and making the wings remain upright. We'll solve them.

I should add that there are two styles of Comparadun wing: 180-degree and deep-V. Viewed from the front, the 180-degree type forms a semicircle, with the bottom flush to the water's surface on both sides. The deep-V forms a very wide V-shape. I like it the better of the two, but both work. What doesn't work is an upright wing: It off-balances the fly, causing it to lie on one side or the other.

Comparaduns are tied with various types of tails, all of which have their place. Here, I want to introduce you to a simple yet very effective tail that the guys at Blue Ribbon Flies in West Yellowstone, Montana, originated. They call flies tied thusly Sparkleduns, because the tail is made from a sparkly synthetic yarn. From below, it resembles the shuck of an escaping insect. Trout seem to key on this, apparently aware that the bug is at their mercy at that moment.

This is another generic sort of fly, and it can be adapted to imitate practically any small-to-medium-sized mayfly. Here, we'll tie one that imitates the prolific western pale morning dun.

***The Dressing*** Western Pale Morning Dun (PMD) Sparkledun

*Hook:* Daiichi #1180 or comparable.

*Hook Size:* 16 or 18; for this exercise, size 16.

*Thread:* 8/0 Uni-Thread or comparable; olive or yellow.

*Wing:* Very fine deer body hair or calf body hair; light gray .

*Tail:* Sparkly yarn—Antron, Z-Lon, or whatever; amber.

*Body/Thorax:* Fine-packing dubbing; pale olivish yellow.

## Tying Steps

1. Tie on and wrap a substantial thread base.
2. Cut off a small bunch of hair, clean it thoroughly, and stack the tips. Not too much hair; this will cause grief at the tying table and poor performance astream.
3. Hold the bunch in your left hand and gauge length. Be conservative; it should be no longer than the shank.
4. Tie the bunch in place about 30 percent of the shank length rearward of the eye. After securing it with some firm wraps, take a look at it from the front. At this point, the hairs should be evenly distributed in a semicircle, with the bottom a very wide V. If the bottom is flat, that's okay; the V will form itself when the thorax is wrapped. If the hair needs to be redistributed, do so with your fingers.
5. Crimp the bunch upright with your right thumb; then stand the hairs up with lots of thread wraps at the base. Use more thread than you would for a hackled fly. Actually, the buildup here approximates that behind the wing.

***Wing hair in place, butts bound down; steps 2-5.***

***Wing formed, yarn tail in place; step 8.***

6. *Option:* If you find that you have problems with the wings falling forward later, do this: Be sure that the wing is shaped as you wish. Then, with a toothpick, put a very small drop of Zap-A-Gap at the base of the hair, and rub it, so that it drys quickly.
7. Trim the wing butts on a long slope. Then run the thread to the bend and tie on a fairly sparse bunch of tailing material, equal in length to the shank. Dovetail the excess with the wing butts and neatly bind it down, ending up back at the bend.
8. The body and thorax can be tied in one continuous operation, provided that you've accurately estimated the amount of dubbing. Spin it on single-thread style and not too thick.
9. Wrap the dubbing to the wing, but don't let it push against the wing from the rear. Come underneath the wing and take the first wrap or two of dubbing very tight to

the wing base. The dubbed thread is now positioned just ahead of the wing on the far side of the hook. Make an X-wrap across the bottom of the hook, coming underneath, up behind the wing, over behind the wing, then under again, coming up in front of the wing. As I stated earlier, don't let the dubbing crowd the back of the wing.

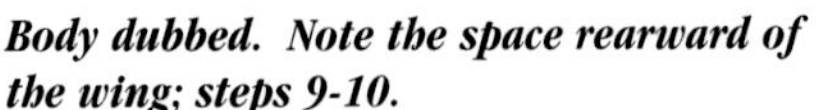

***Body dubbed. Note the space rearward of the wing; steps 9-10.***

***Thorax being dubbed; step 10.***

**10.** Finish off the shaping of the thorax, whip-finish, lacquer, and your Compara/Sparkledun is complete.

***Two views of the completed PMD Sparkledun.***

## The Down-Wing

This name is a bit misleading; it might make you think the wing is made out of a downy material of some sort. Actually, the reference is to the position of the wing, which lies more or less prone above the body rather than upright. Quite a few insects that are of importance to the angler carry their wings in this position, including the caddis flies, the stone flies, and various terrestrials.

One of the most popular and successful down-wing flies is the Troth Caddis, which was originated by Al Troth, who left his native Pennsylvania to become a legendary guide and fly tyer on Montana's Beaverhead River. I'm not sure what Al's original dressing looked like, because I've seen several versions all bearing the same name.

The one I want to show you here is quite basic: body, hackle, and wing. The key ingredient is the pale-colored deer or elk hair that forms the wing; it's light and buoyant. In smaller sizes, deer body hair is the better choice. You can adapt this design to practically any species of caddis as long as the adult is of sufficient size for hair to be used as a wing imitation.

A further word about hair for down-wings. The body hair from deer, elk, and other such animals is quite different than the bucktail you used earlier in this book. It's softer and has a tendency to crimp and flare. These characteristics vary considerably, though, depending on the specific animal and the part of the body from which the hair came.

For spun-and-trimmed-hair procedures, such as those used for bass bugs and Muddler Minnow heads, hair that flares readily is desirable. For down-wings, a moderate amount is fine, but too much flaring will result in a rather wild-looking wing. The portion toward the tips flares less than that toward the butts, so the outer one-third of a bunch of hair will usually work pretty well.

### *About Hackling Down-Wing Flies*

As you've learned, the standard hackle length on a dry fly is 1½ times the gape of the hook. For the type of fly we're about to tie, that's a bit long—the result would be a fly that rode too high and didn't balance well on the water. For best results, the *palmered* (spiral-wrapped over the body) hackle should extend approximately to the point of the hook.

With the quality of today's hackle, especially saddles, finding feathers that will accommodate these proportions shouldn't be a problem. However, if your hackles come out a bit long, it's allowable to trim the bottoms just a tad.

***The Dressing*** Hair-Wing Caddis

*Hook:* Daiichi #1180 or comparable.

*Hook Size:* For this exercise, size 12 or 14.

*Thread:* 8/0 Uni-Thread or comparable; brown or tan.

*Hackle:* Tan or ginger-grizzly, preferably saddle, wrapped over the body.

*Body:* Green stretch nylon or dry-fly grade dubbing.

*Wing:* Light-colored elk or deer hair.

#### Tying Steps

1. Start the thread just a bit to the rear of the eye and wrap to the bend. There, tie in a hackle feather that you've prepared in accordance with previous instructions. Remember to leave that tiny length of quill showing. Run the thread forward and back, binding down the quill, and trim off the excess. The thread ends up back at the bend.

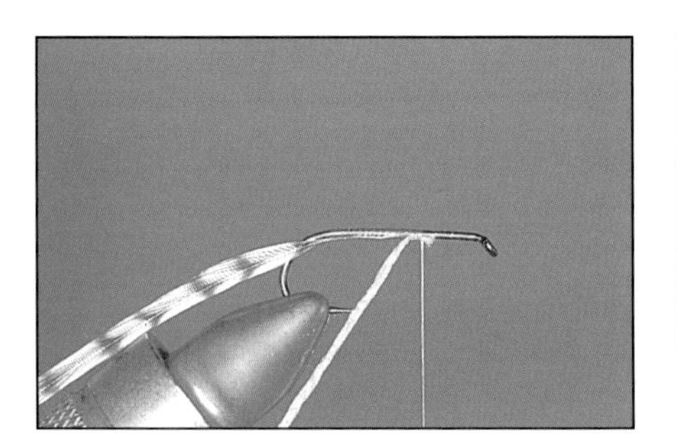

***Hackle and body material tied in; step 1.***

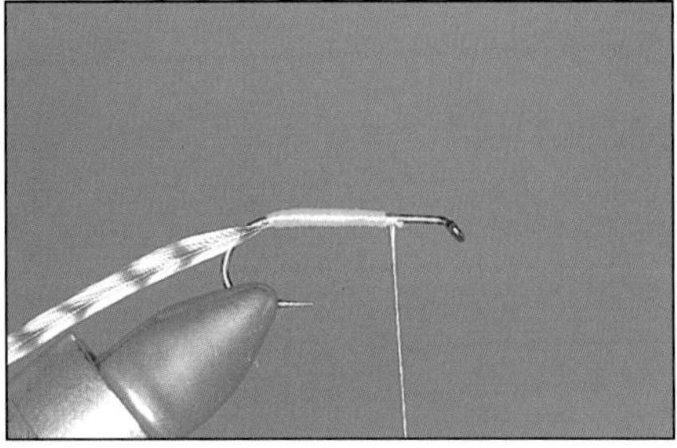

***Body completed; step 2.***

2. Make a slender body, stopping 25 to 30 percent of the shank length to the rear of the eye, as shown.
3. Spiral-wrap the hackle and secure it at the front of the body with a few turns of thread. *Don't* cut off the remainder; just let it hang.

4. Prepare for the wing by making a two- or three-layer thread base between the eye of the hook and the front of the body. The thread ends up near the rear of this base, a couple of turns ahead of the body. Also, with the tips of your scissors, cut a channel into the hackle on top of the hook, where the wing will go. From the front, this would look like a wide V.

***Hackle wrapped and secure, thread base for wing in place; steps 3-4.***

***Cutaway V to accommodate wing, front view; step 4.***

5. Clean out and stack a small bunch of the pale hair. When compressed, it should be about the diameter of a wooden kitchen match.
6. Hold the bunch with your left thumb and forefinger and set it on top of the hook, in the channel you trimmed out. It should be slightly longer than the body; check the photo for proportions.
7. Tie on the hair, keeping it centered on top of the hook. For the first few wraps, apply only moderate thread tension, so as not to cause the hair to flare too much. Then secure the wing with a series of very firm wraps.
8. When the wing is absolutely secure, trim the butts neatly and bury them under a smooth layer of thread, thus creating a base for the front hackle.

***Wing in place, new thread base wrapped to accept front hackle; steps 5-8.***

***The completed Hair-Wing Caddis.***

9. Regrip the feather and make a front hackle; these wraps are contiguous, not spiraled. Tie a whip-finish, coat with lacquer, and go find a caddis emergence.

The down-wing is quite easy to tie, and once you've mastered it, a wide array of great patterns will be within your capabilities. In the final chapter, I'll list the dressings for several that have become favorites of mine.

## The Spent-Wing

For our last exercise, I want to show you the spent-wing, which is used to imitate the adult, or imago, form of the mayfly. After their final molt, these shimmering, semitransparent insects are ready for their glorious culminating moment: the mating dance. As evening falls, they come forth from the streamside foliage and begin their nuptials. The males and females join in midair, and the eggs are instantly fertilized. The females then zoom over the riffles, depositing their precious cargo, ensuring continuity of the species. At last they drop to the water, utterly spent, to become nutrients for the trout waiting expectantly in the pool below. Human life should be so efficient.

There are several methods for making this type of wing. Synthetic yarns and fibers have become quite popular, and they do work, to a degree. My main complaint about them is that they don't hold their shape very well under fishing conditions. For this reason, I've gone back to my hackle spent-wings, with no regrets.

Besides the wings, you'll also learn how to make realistic tails. These can be fashioned out of long, stiff hackle barbs, but I prefer a synthetic product known as Microfibetts. (I don't know the derivation of this term.) They're sold in small envelopes and resemble nylon paintbrush bristles. Many colors are available, but for spent tails, a few shades of gray will pretty much cover everything.

I might add that if you find tying realistic tails tedious, don't bother with them. Just tie on a sparse bunch of long hackle barbs, well spread, and go fishing.

### The Dressing Rusty Spinner

*Hook:* Daiichi #1180 or comparable.

*Hook Size:* Various; for this exercise, size 12.

*Thread:* Tan, or light or medium brown.

*Wings:* Slightly oversized hackle; pale silvery gray.

*Tails:* Two or three Microfibetts (see note); silvery gray.

*Body/Thorax:* Bright rust-colored or fiery brown dubbing.

**Note:** Some mayfly spents have two tails, some have three. Here, we'll tie three. To make a two-tailed spent fly, follow the same process, and simply omit the center tail.

### Tying Steps

1. Tie on near the front, wrap a thread base, and tie in one large hackle (after preparing it in the usual manner). It should be one or possibly two sizes larger than would be normal for this hook size. In selecting this feather, look for the thinnest possible quill.

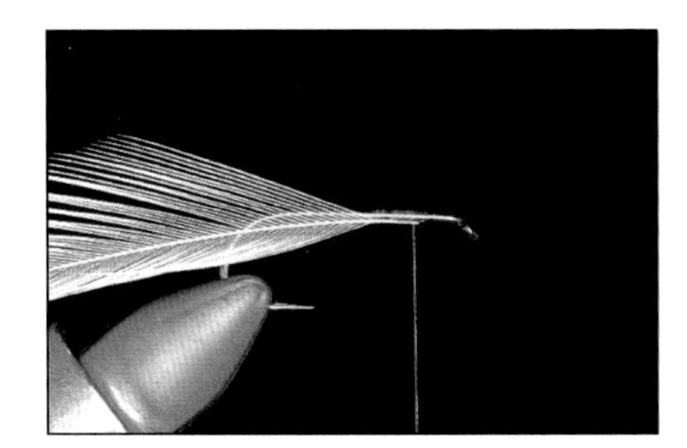

*Oversized hackle feather tied in; step 1.*

2. This hackle is wrapped in an unconventional manner. Instead of progressing forward after the first two turns, begin making crisscross wraps, fore and aft. If the barbs tend to flare, don't worry; that's just what you want.

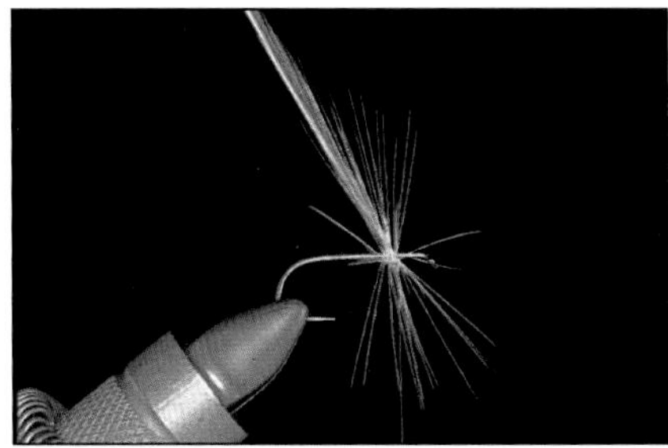

*Crisscross-wrapping the hackle; step 2.*

***Crisscross-wrapping the hackle; step 2.***

3. When you've finished wrapping the hackle, tie it off in front. Then take your sharp-pointed scissors and cut a V top and bottom. Not too wide a V; you don't want to sacrifice too much material. The dubbing that you'll apply in a moment will take care of gathering and flattening out the barbs in spent-wing fashion.
4. Pass the thread through the bottom V and take a few wraps behind the wings. Then tie three Microfibetts on top of the hook, gauging the proportions to be extra long. Hold them with your left thumb and forefinger and wrap to the bend; trim the butts. The reason for tying on the tails in this manner is to avoid having to wrap forward and back to bury the butts (which is what you'd need to do if you tied on the tails at the rear).
5. Bend the Microfibetts on each side outward, crimping them at an angle to the shank.
6. Pass the thread over the hook and between the center and outermost Microfibett, using your left forefinger to keep the two separated. Jam the thread into the notch, so that the outer fiber is supported at about a 45-degree angle.

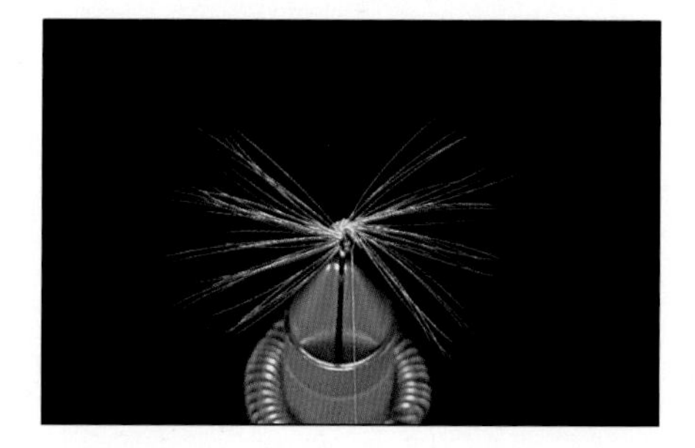

***Top and bottom V-cuts; step 3.***

7. Bring the thread under the hook. Hold the nearest fiber firmly and pass the thread between it and the center tail fiber, again forcing it into the notch. This is a left-handed move; you grip the tail fiber with your right hand, and pass the thread through with your left. Then take one thread wrap adjacent to the base of the tail.

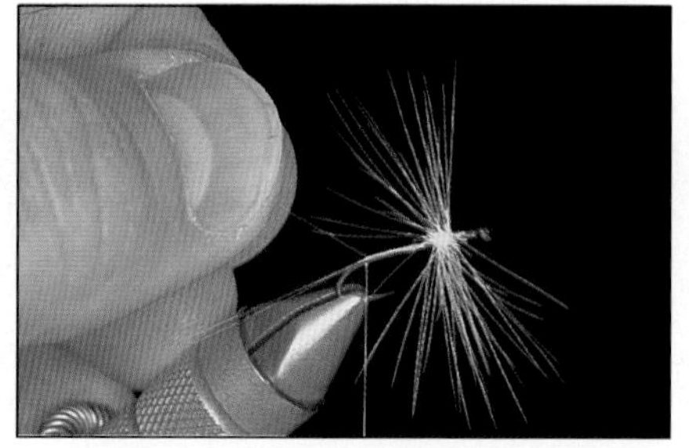

***The tailing process; steps 4-7.***

8. Spin on a thin worm of dubbing. If you can estimate what you'll need to form the body and thorax, top and bottom, fine. If you're unsure, be conservative; you can always add more as you go. In the photo, I'm using about 3 inches (75mm) of dubbing.
9. *Option:* If you'd like to further isolate the tail fibers, use the beginning of the dubbing to repeat the separating moves you just made with the thread. It'll go easier, since the tails are already well spread. They'll remain in place anyway, but this technique does create a nice effect at the rear of the body.
10. Wrap the body. When you reach the wing, come underneath, through the bottom V. Wrap once around the hook then back through the bottom V, creating an X of dubbing. Do the same on top, X-wrapping the dubbing through the upper V. With very finely packed dubbing, you can repeat this procedure, making two X-wraps top and bottom, but don't overdo. End up in front, whip-finish, and, if you wish, tint the thread with a brown marker. Apply lacquer, go find a quiet pool below a riffle, and wait for dusk.

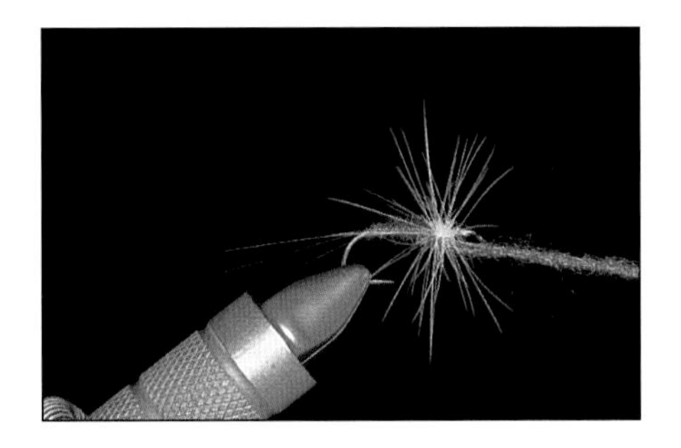

***Dub the body, then come underneath, through the bottom V; step 10.***

***Crisscross-wraps bottom and top, behind and in front of wings; step 10.***

***The completed Rusty Spinner.***

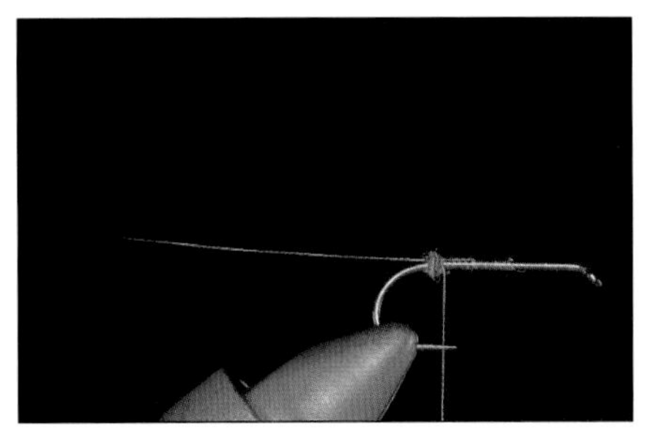

***Dividing the tails with a dubbing ball.***

### *Tailing Option*

Many tyers prefer a different method for making a split tail. Here's the procedure:

1. Instead of tying on the tailing fibers just behind the wing, wrap to the rear, and wrap a tiny, tightly packed ball of dubbing right at the bend.
2. On each side of the hook at a time, tie a tailing fiber, and wrap the thread rearward, forcing the fibers against the ball of dubbing.
3. This creates a twin split tail. If you want three tails, simply tie in a center fiber before applying the dubbing.

This process is also used for making the prominent "outrigger" tails that are required to balance the no-hackle type of fly. Small bunches of barbs replace the Microfibetts.

As I mentioned at the beginning of this chapter, there are still more methods of winging dry flies. There's the upright post, which accommodates parachute hackle; the duck-quill-section wing, which is ancient and traditional indeed; the no-hackle wing, which is also made of duck or goose wing feather sections, but is tied differently; and others. They all have their place—but in my view, it's not in a beginner's book.

SEVEN

# *Putting It All to Work*

If you can tie the flies in this book with reasonable competence, you've made a terrific start. To be sure, there are other techniques to be learned and refinements to be implemented. It's a matter of just how much you want to achieve to satisfy yourself as a fly tyer and to support your angling needs.

The thing to remember is that similar materials are all handled in the same manner, or virtually so. For example, hair is handled like hair, be it bucktail, squirrel tail, bear, goat, synthetic, or whatever. Certainly, there are differences; all materials have their idiosyncracies. But with the understanding and the skills that you now have, you can adapt and cope.

You also now have a fairly comprehensive familiarity with the language of fly tying. You know what *thread torque* means and the difference between a soft-wrap and a firm one. You understand the pinch- and the distribution-wrap. You know what a hackle barb is. You understand how to grade hackle. You've learned a great deal, and now you can put it to work for you.

There are lots of pattern books available today. Some are very broad in their approach, and some are style-specific. Some have lots of new stuff, whereas others catalog the traditional. I think you'll be pleasantly surprised when you start looking through the literature of fly tying and realize you can say to yourself, "Hey, I can tie that!"

There are also several excellent periodicals for the fly tyer. These are of particular value, because they'll keep you informed as to what's new—and I assure you there will be plenty of developments and innovations. Fly tying today is very dynamic; things keep moving along, spurred by both the creative juices of the tyers themselves and the need to develop new materials to replace those no longer available. The field of Atlantic salmon fly tying is a prime example. Modern high-tech materials are quickly taking over, replacing the rare and exotic plumage that was so much in vogue during the days of the stained-glass-window school of fly tying but is now illegal, unavailable, or both.

What it all boils down to is this: mixing and matching. For example, just consider how many wet flies and nymphs have either dubbed, floss, yarn, chenille, or herl bodies. With a few exceptions, the differences are in the wings and hackle and perhaps the ribbing. If you've mastered the procedures in chapters 3 and 4, a large percentage of the dressings you'll encounter are now in your repertoire—and that includes the less-complicated salmon and saltwater patterns.

In my opinion, the main challenge facing fly tyers today is selection of materials, especially new materials. All you have to do is attend one of the major consumer shows that take place throughout the country every winter to see what I mean. Some of this stuff is wonderful and has broadened the scope of fly tying. But there's also some mystery stuff out there that I find to be of questionable value, at best. In some cases, I don't believe the new materials have been tried out by competent tyers and anglers. You'll have to watch out for these sorts of things.

With that admonition, I wish to present to you a selection of various types of dressings that I find particularly useful and attractive. This isn't a long list—but then, this isn't a pattern book, per se. I do believe that you'll enjoy tying these patterns and that they'll serve you well astream.

In order to simplify matters and avoid a lot of tedious redundancy, I'm not going to specify the hooks and threads as I did for the patterns in the preceding chapters, except in the case of materials that are not covered earlier, not obvious from the photographs, or very specific. It should be an easy matter for you to select the appropriate hook and thread: Just look at the photo of the fly you're about to tie, then cross-relate it to a similar one in the book. As to thread color, it should simply be compatible with the body color. If necessary, you can anoint the head with a felt marker.

There's one thing I do want to emphasize: Some of these flies are tied in a wide range of sizes. For example, I tie Wulffs from size 8 to 18. It stands to reason that I'm not going to use exactly the same material on, say, a 16 that I used on an 8. The colors and characteristics will be quite similar, but I'll make substitutions, as appropriate, to accommodate hook size. This applies to body, winging, and tailing materials in particular. Please do likewise.

## Dry Flies

### *Eastern March Brown*

*Wings:* Barred teal or mallard flanks.

*Tail:* Mixed dark brown and grizzly.

*Body:* Creamish dubbing.

*Ribbing:* Flat, fairly wide brown thread, tied in at the front of the body and wrapped rearward and forward to form a diamond pattern.

*Hackle:* Mixed brown and grizzly.

*Comments:* This traditional dressing imitates a famous northeastern mayfly, *Stenonema vicarium.* Tied in a pale shade, it also imitates a closely related mayfly, *Stenonema fuscum,* commonly called the gray fox. Sizes range from 8 to 12.

### *Royal Wulff*

*Wings:* White hair: calf tail, etc.

*Tail:* Brown hair or hackle barbs.

*Body:* Royal Coachman style; two bumps of peacock, separated by a band of bright red flosslike material, such as stretch nylon.

*Hackle:* Dark brown.

### *Grizzly Wulff*

*Wings:* Brown hair: calf tail, etc.

*Tail:* Same as wings.

*Body:* Yellow dubbing, yarn, or similar body-wrap.

*Hackle:* Mixed brown and grizzly.

### *White Wulff*

*Wings:* White hair: calf tail, etc.

*Tail:* Same as wings.

*Body:* White or off-white dubbing or yarn.

*Hackle:* Light badger.

*Comments:* I'll deal with the Wulffs as a group. They're fantastic general patterns. They were designed for more diffused currents, and that's where they work best, but on occasion they'll take fish on quiet waters if there's insect activity of a sort that the particular Wulff pattern imitates reasonably well. As of this writing, there are probably several dozen variations of Lee Wulff's enduring design. Sizes typically range from 10 to 14, but the flies can be tied both larger and smaller than that.

### *House & Lot (H&L Variant)*

*Wings and Tail:* Wulff style; white.

*Body:* Rear half is stripped peacock; front half is peacock herl.

*Hackle:* Furnace or brown.

*Comments:* This is a wonderful Rocky Mountain fly. The original dressing called for a ribbing of fine gold wire. Instead, I Zap-A-Gap the quill portion of the body. In contrast to the Quill Gordon, this body utilizes a frond from the peacock tail proper. With a little practice, you can make the quill and herl portions of the body out of one frond. The usual size range is 10 to 14.

### *Dun Variant*

*Tail:* Dark gray hackle barbs.

*Body:* Stripped hackle quill from a dark brown cape.

*Hackle:* Dark gray.

### *Gray Fox Variant*

*Tail:* Mixed ginger and grizzly hackle barbs.

*Body:* Stripped hackle quill from a ginger cape.

*Hackle:* Barred ginger and grizzly.

*Comments:* These flies are tied in similar fashion to the Cream Variant in the book. The original dressing for the Gray Fox Variant called for light ginger, dark ginger, and grizzly hackles. The same effect is obtained more easily by using the mix specified above. Sizes range from 12 to 16.

### *Light (Pink) Hendrickson*

*Wings:* Lemon-barred wood duck flank.

*Tail:* Medium-to-light gray hackle barbs.

*Body:* Pinkish-lavender dubbing.

*Hackle:* Medium-to-light gray.

*Comments:* This is an Art Flick pattern. It was designed to imitate the female dun of the mayfly, *Ephemerella subvaria.* The original dubbing was urine-burned vixen fox fur. Today, synthetics and mixes are available that successfully emulate the elusive shade. It is usually tied in size 12.

### *Red Quill*

*Wings:* Lemon-barred wood duck flank.

*Tail:* Medium gray hackle barbs.

*Body:* Stripped quill from a brown rooster cape.

*Hackle:* Medium gray.

*Comments:* Art Flick being a classic Catskill gentleman, he would never think of leaving a fair lady unescorted; thus this pattern that imitates the male dun of *E. subvaria.* It is usually tied in size 12 or 14.

### *Quill Gordon*

*Wings:* Lemon-barred wood duck flank.

*Tail:* Medium gray hackle barbs.

*Body:* Stripped quill from the "eye" of a peacock tail feather.

*Hackle:* Medium gray.

*Comments:* This is a great American classic. The original dressing is attributed to the immortal Theodore Gordon and was allegedly designed to imitate an early-season northeastern mayfly, *Epeorus pluralis.* Again, I have replaced a fine gold wire rib with Zap-a-Gap. It is usually tied in sizes 12 and 14.

### *Adams*

*Wings:* Grizzly hackle tippets.

*Tail:* Mixed brown and grizzly

*Body:* Fine gray dubbing.

**Dry Flies, *Left to Right, From the Top:***
***Row 1: Royal Wulff, Grizzly Wulff, White Wulff***
***Row 2: Western Green Drake, H&L Variant, Eastern March Brown***
***Row 3: Quill Gordon, Dun Variant, Gray Fox Variant***
***Row 4: Pink Hendrickson, Lime Trude, Red Quill***
Photograph by Tim Savard.

*Hackle:* Mixed brown and grizzly.

*Comments:* With the possible exception of the Royal Coachman, this is the world's most ubiquitous dry fly. It is usually tied in sizes 12 to 18.

### *Light Cahill*

*Wings:* Lemon-barred wood duck flank.

*Tail:* Straw-cream hackle barbs.

*Body:* Creamish dubbing.

*Hackle:* Straw-cream.

*Comments:* This is another old classic that imitates a number of mayflies and is a good general pattern as well. It is usually tied in sizes 12 to 16.

### *Tan/Grizzly Caddis*

*Hackle:* Grizzly saddle, palmered over body, then wrapped in regular fashion ahead of the wing.

*Body:* Tannish dubbing.

*Wing:* Fine deer hair.

*Comments:* The caddis design in the book proper can be used to imitate any adult caddis you wish. This is a case in point. It is tied in a wide range of sizes.

### *Henryville Special*

*Hackle:* Grizzly saddle, palmered over body.

*Body:* Pale olive or bright green flosslike material.

*Underwing:* A small bunch of lemon-barred wood duck.

*Wings:* Gray duck or goose wing feather slips, tied so that they flare away from each other, tips pointing downward.

*Front Hackle:* Brown.

*Comments:* This is a slightly more sophisticated caddis pattern that has been a favorite for years. It is also tied in a wide range of sizes.

### *Hemingway Caddis*

*Palmered Hackle:* Smoky gray saddle.

*Body:* Olive dubbing; about 65% of the shank.

*Underwing and Wing:* Same as for the Henryville.

*Thorax:* Peacock herl.

*Front Hackle:* Same as palmered hackle, wrapped through the thorax.

*Comments:* This dressing is that of Jack Hemingway, son of the redoubtable Ernest. It is a terrific pattern for imitating several caddis found in the Rocky Mountain watersheds. It is also tied in a wide range of sizes.

### *Sierra Red Dot*

*Tail:* Golden-pheasant tippet, in strands.

*Hackle:* Grizzly, rear and front.

*Belly Band:* Hot red flosslike material, such as Chinese red Uni Nylon Stretch.

*Comments:* This is a California pattern that works everywhere. For variety, you can use different-colored belly bands. It is commonly tied in sizes 12 to 18.

### *Brown Bivisible*

*Main Hackle:* Brown.

*Front Hackle:* White.

*Comments:* This is another color of the basic fly from chapter 6 in sizes 10 to 16.

### *Lime Trude*

*Tail:* Golden pheasant tippet, in strands.

*Body:* Lime-colored flosslike material; about 2/3 of the shank.

*Wing:* White hair, tied caddis style.

*Hackle:* Mixed brown and grizzly.

*Comments:* The Trude style represents a generic series. There are a number of color combinations; this one is a proven killer. It is tied in sizes 10 to 16.

### *Pale Morning Dun*

*Wings:* Pale gray hen tippets.

*Tail:* Pale gray hackle barbs.

*Body:* Light yellowish olive dubbing.

*Hackle:* Pale gray.

*Comments:* This pattern imitates a ubiquitous western mayfly. I've noticed that the body color varies a bit from stream to stream. It is tied in sizes 16 to 18.

### *Dorato Hare's Ear*

*Wings:* Lemon-barred wood duck flank.

*Tail:* Mixed brown and grizzly, tied very short.

*Body:* Hare's-ear dubbing.

*Hackle:* Mixed brown and grizzly, trimmed a little on the bottom.

*Comments:* This unlikely looking creation of Bill Dorato of Albany, New York, is one of the great all-around patterns. It's tied in sizes 10 to 16.

***Dorato Hare's Ear, Light***

*Wings:* Lemon-barred wood duck flank.

*Tail:* Mixed brown and straw cream, tied very short.

*Body:* Pale hare's-ear dubbing.

*Hackle:* Mixed brown and straw cream, trimmed a little on the bottom.

*Comments:* This is a pale-colored version of the DHE.

***Griffith Gnat***

*Hackle:* Grizzly, preferably saddle.

*Body:* Peacock herl.

*Comments:* This is a simple yet deadly dressing that covers the adult midge situation very well indeed. It is usually tied small to very small.

***Western Green Drake***

*Wings:* Smoky gray deer hair; very fine.

*Tail:* Gray-dyed grizzly hackle barbs.

*Body:* Bright olive dubbing.

*Ribbing:* Wide, flat, yellow thread, tied on at the front of the body and wrapped to the rear and back to form a diamond pattern.

*Hackle:* Gray-dyed grizzly.

*Comments:* This dressing covers the several mayflies that make up the western green drake group. It's commonly tied in sizes 8 to 14.

***Callibaetis Sparkledun***

*Wing:* Speckled deer hair, fine.

*Tail:* Yarn—Antron, Z-lon, or whatever; dark olive-gray .

*Body/Thorax:* Creamy tan dubbing.

*Comments:* This dressing imitates the more common *Callibaetis* found throughout the West. It's generally tied in sizes from 16 to 20. The colors can be varied to suit local hatches.

**Dry Flies,** ***Left to Right, From the Top:***
***Row 1: Dorato Hare's Ear, Dark, Adams, Light Cahill***
***Row 2: Dorato Hare's Ear, Light, Sierra Red Dot, Olive Sparkledun***
***Row 3: Pale Morning Dun, Griffith Gnat, Brown Bivisible***
***Row 4: Henryville Special, Hemmingway Caddis, Hair-Winged Caddis-Grizzly***
Photograph by Tim Savard.

## Wet Flies

### *Picket Pin*

*Tail:* Brown hackle barbs.

*Hackle:* Brown; palmered.

*Body:* Peacock herl.

*Wing:* Gray-squirrel tail.

*Head:* More peacock; two wraps.

*Comments:* I don't think anyone is sure what this fly is supposed to represent, but it certainly is a fish-catcher. It's tied from sizes 6 to 14 on hooks ranging from 2XL to 4XL.

### *Leadwing Coachman*

*Tag:* Gold flat tinsel; just a couple of turns.

*Body:* Peacock herl.

*Hackle (Beard):* Brown.

*Wings:* Slate duck or goose wing feather slips.

*Comments:* One of the all-time killers. It's commonly tied in sizes 10 to 16.

### *Grannom*

*Tag:* A little ball of fine pea-green chenille or dubbing.

*Ribbing (optional):* Fine gold oval tinsel.

*Body:* Tannish dubbing or yarn.

*Hackle (Beard):* Hungarian partridge.

*Wings:* Slips from mottled turkey wing feathers.

*Comments:* This is a very useful fly when the caddis are diving into the water to oviposit, which certain species do. A number of different dressings go by this name. Commonly tied in sizes 10 to 14.

### *Cow Dung*

*Tail:* Brown hackle barbs.

*Ribbing:* Flat, narrow gold tinsel.

*Body:* Pale olive floss.

*Hackle (Beard):* Soft, brown hackle barbs.

*Wings:* Brown.

*Comments:* While the name may seem unappetizing to us, the trout don't seem to agree. It's an old English pattern, commonly tied in sizes 10 to 16.

### *Alder*

*Tag and Body:* Same as for Leadwing Coachman.

*Hackle (Beard):* Black, soft.

*Wings:* Mottled turkey, like the Grannom.

*Comments:* This is a good all-around wet fly that also imitates a couple of northeastern insects known as alder flies. Tie it in sizes 10 to 14.

### *Gold-Ribbed Hare's Ear*

*Tail:* Hungarian partridge.

*Ribbing:* Narrow, oval gold tinsel.

*Body:* Hare's-ear dubbing.

*Throat:* Hungarian partridge.

*Wings:* Same as for Leadwing Coachman.

*Comments:* Arguably the most successful of the traditional wets, this revered dressing originated across the pond. It's tied in sizes 10 to 16.

### *Grizzly King*

*Tail:* Red-dyed hackle barbs.

*Ribbing:* Narrow, flat silver tinsel.

*Body:* Bright green floss or substitute.

*Hackle (Beard):* Soft grizzly.

*Wing:* Mallard or teal flank.

*Comments:* This is a cousin of the Professor from the wet-fly chapter. It's a fine attractor pattern commonly tied in sizes 8 to 14.

### *Montreal*

*Tail:* Claret-dyed hackle barbs.

*Ribbing:* Narrow, flat gold tinsel.

*Body:* Claret floss or substitute.

*Hackle (Beard):* Claret-dyed hackle barbs, soft.

*Wing:* Mottled turkey à la the Alder.

*Comments:* This is another old favorite and a handsome one that's tied in sizes 10 to 14.

### *Grouse & Green Soft-Hackle*

*Ribbing:* Narrow gold tinsel, oval or flat.

*Body:* Green flosslike material.

*Hackle (Collar):* A couple of turns of Hungarian partridge.

*Comments:* One of a very large family of soft-hackle wet flies, this is a longtime favorite of mine. Tie it in sizes 10 to 14.

### *Ellis' Fly*

*Tail:* A few wisps of barred wood duck or gray-phase Hungarian partridge.

*Body:* A stripped quill from the eye of a peacock tail feather.

*Thorax:* Peacock herl; just a small bump.

*Hackle (Collar):* A couple of turns of gray-phase Hungarian partridge.

*Comments:* This fly was shown to me by Mr. Ellis Hatch of Rochester, New Hampshire. Also a member of the soft-hackle family, it works wonders as an emerging-midge pattern, especially in still waters. It's tied in a wide range of sizes, commonly from 10 to 16.

### *Brown-Hackle Peacock*

*Body:* Peacock herl.

*Hackle (Collar):* A few turns of soft brown hackle.

### *Grizzly-Hackle Peacock*

*Body:* Peacock herl.

*Hackle (Collar):* A few turns of soft grizzly hackle.

*Comments:* These two dressings are also soft-hackles. Simple as they are, they're nevertheless productive and can be tied in a couple of minutes. Both are sometimes dressed with a very short red yarn tag and/or oval silver tinsel ribbing. Tie them in sizes 10 to 16.

## Nymphs

### *Zug Bug*

*Tail:* Three peacock sword barbs.

*Ribbing:* Oval silver tinsel.

*Body:* Peacock herl.

*Hackle (Beard):* Brown, soft.

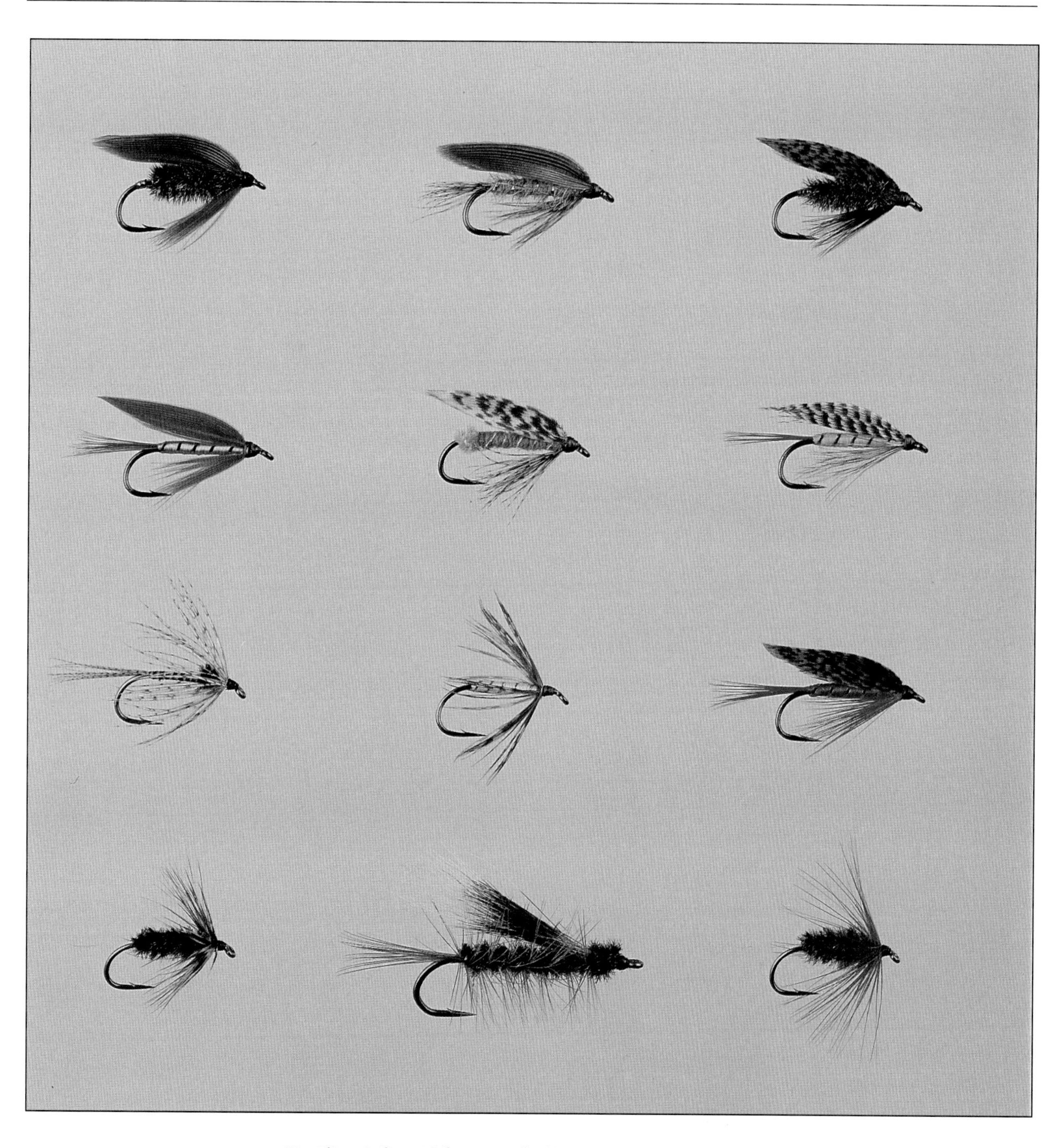

**Wet Flies,** ***Left to Right, From the Top:***
***Row 1: Leadwing Coachman, Gold-Ribbed Hare's Ear, Alder Fly***
***Row 2: Cow Dung, Grannom, Grizzly King***
***Row 3: Ellis' Fly, Grouse & Green Soft-Hackle, Montreal***
***Row 4: Grizzly-Hackle Peacock, Picket Pin, Brown-Hackle Peacock*** Photograph by Tim Savard.

*Wing Case:* A small piece of mallard, teal, or wood duck flank feather, trimmed to shape.

*Comments:* This is an excellent general nymph pattern that's tied in sizes 8 to 14.

### *Gray Nymph*

*Tail:* Dark gray hackle barbs.

*Underbody (optional):* Fine lead wire.

*Body/Thorax:* Gray dubbing.

*Wing Case:* Dark brown or gray Bugskin, or a strip of sprayed feather.

*Hackle (Beard):* Dark gray hackle barbs.

*Comments:* This is another fine all-around dressing that's tied from sizes 8 to 18.

### *Talleur's Golden Stone*

*Tail:* Two medium brown biots tied à la the Prince Nymph.

*Underbody:* Lead wire; thickness according to the size of the nymph you're tying.

*Ribbing:* Brownish Maxima monofilament; thickness according to fly size.

*Body/Thorax:* Amberish gold dubbing; Scintilla #63 is very nice.

*Hackle:* Brown webby saddle feather.

*Wing Case:* Bugskin, mottled brown.

*Comments:* This is a simple yet effective pattern of mine that covers the numerous *Perlidae* stone flies, east and west. It is tied in sizes 4 to 12.

### *Tellico*

*Tail:* Brown hackle barbs.

*Ribbing:* Peacock herl.

*Wing Case:* Either strands of peacock herl or cock pheasant tail fibers, covering the entire top of the body.

*Body:* Deep yellow floss or substitute, tied full and tapered.

*Hackle:* Brown.

*Comments:* This is a rather unusual looking but consistently effective pattern. Note that both the ribbing and the wing case material must be tied in at the rear before the body is formed. It's commonly tied in sizes 10 to 14.

### *Whitlock Red Squirrel Nymph*

*Tail:* Orangey fur or hair from a fox-squirrel pelt or tail.

*Rib:* Oval tinsel or wire, gold or copper.

*Underbody:* Lead wire.

*Body:* Orangey amber fur from a fox-squirrel pelt, with some sparkly synthetic dubbing of a similar shade mixed in.

*Thorax:* Fur, including guard hairs, from the dark part of a fox-squirrel pelt; the best is along the spine.

*Hackle (Collar):* Hungarian partridge or brownish speckled hen saddle.

*Comments:* This is a wonderful dressing from the creative hands of the great Dave Whitlock. It is tied in sizes 6 to 14.

### *Isonychia*

*Tail:* Dark brownish partridge or speckled hen barbs.

*Stripe (optional):* Heavy white thread.

*Shellback/Wing Case:* A strip of very dark brown or gray Bugskin.

*Ribbing:* Smoky gray or brown monofilament; Maxima 5X works fine.

*Body/Thorax:* Dark brownish dubbing; Ligas Brown Drake is ideal.

*Hackle (Legs):* Same as tail, tied short.

*Comments:* The genus *Isonychia* has many important members, and many patterns have been devised to imitate them. This one is mine. If you opt for the white stripe, which runs down the center of the back, tie in the thread immediately after making the tail. After folding the Bugskin forward to form the shellback, lay the white thread over the back, fasten it at the thorax with a few wraps of thread, and wrap the ribbing. Trim off the excess ribbing material, but not the white thread. Then, dub in the thorax, tie on the legs, fold forward and tie down the remaining Bugskin, and, lastly, bring the rest of the white thread over the top. This fly is tied in sizes 10 to 14.

### *Olive Drake Nymph*

*Tail:* Dark olive-dyed pheasant tail fibers.

*Ribbing:* Fine oval copper oval tinsel or wire.

*Underbody:* Dark olive dubbing.

*Body:* Dark olive-dyed pheasant tail fibers.

*Thorax:* Same as underbody.

*Wing Case:* Adhesive-treated feather, or Bugskin; dark coppery gray, brown, or olive.

*Hackle (Legs):* Same as tail.

*Comments:* While this dressing was designed to imitate several important western mayfly nymphs, it also serves as a "near-enough" pattern for many other mayflies. It's tied in a very wide range of sizes, including 18 to 22 for the little *Baetis* nymphs.

### *Balloon Emerger*

*Tail:* Hungarian partridge barbs.

*Body/Thorax:* Tannish dubbing.

*Wing Case:* Gray closed-cell foam.

*Hackle (Legs):* Hungarian partridge barbs.

*Comments:* This particular emerger is simply a nymph tied with a foam wing case. Foam of this type is sold in most fly shops. This nymph is commonly tied in sizes 8 to 16.

### Compara-Emerger

*Tail:* Two small bunches of gray hackle barbs, one on each side of a small ball of dubbing.

*Body:* Blended spectrumized fur mixture.

*Wing:* Gray, soft deer hair.

*Head:* The clipped butt ends of the wing.

*Comments:* This is a Caucci-Nastasi pattern. The spectrumized fur, which is used for both the body and the little ball that divides the tails, is sold in fly shops under their names. I tie my tails as follows: Make the divider, then tie on the small bunches of hackle barbs on each side. When the fly has been completed, apply a little head cement—Dave's Flexament is very good for this—to the two bunches, which brings the barbs together. The fly is usually tied in sizes 8 to 16.

### Hendrickson Poly-Fluff Emerger

*Tail:* Barred wood duck.

*Body/Thorax:* Reddish brown dubbing.

*Legs:* Barred wood duck.

*Wing:* Smoky gray Poly-Fluff or substitute.

*Comments:* Poly-Fluff is a product of a company called L&L. If you can't find it, use Z-lon or something like that. For imitating emergers during the eastern or midwestern Hendrickson hatch, tie in sizes 10 to 14.

### PMD Emerger

*Tail (Trailing Shuck):* Amberish or brown Antron-type yarn.

*Body/Thorax:* Yellowish-tannish dubbing, preferably with some sparkle.

*Wing Case:* A ball of gray closed-cell foam.

*Hackle (Legs):* Soft gray material, such as Chickabou.

*Comments:* This pattern imitates the emergent form of one of the West's most prolific mayflies. Chickabou is a registered trade name of Hoffman Hackle, produced by Whiting Farms, Inc. It's the soft stuff from the rump of a saddle. This fly is generally tied in sizes 16 to 18.

### Western Green Drake Emerger

*Tail:* Hungarian partridge barbs.

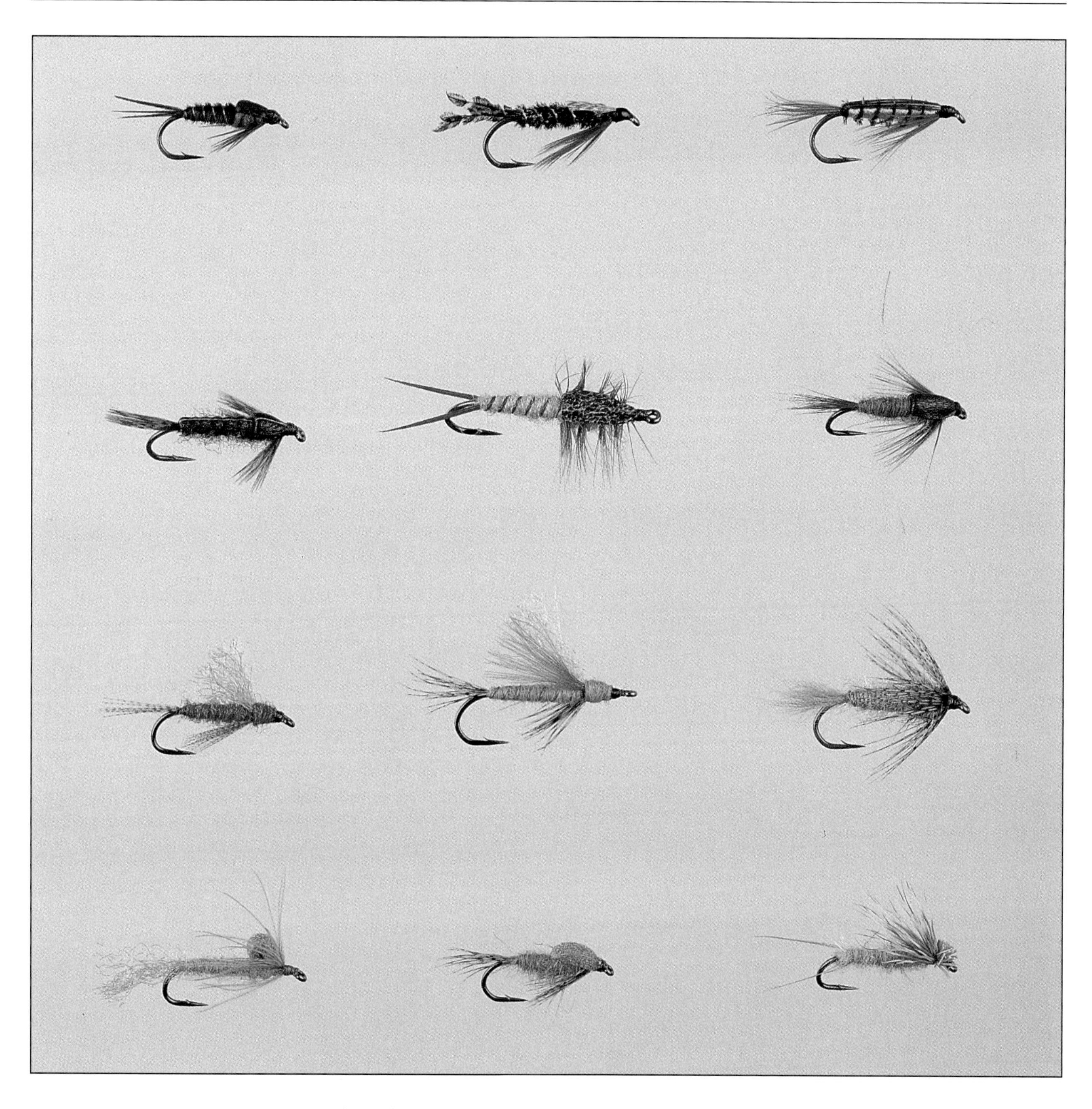

**Nymphs, *Left to Right, From the Top:***
***Row 1: Olive Drake Nymph, Zug Bug, Tellico***
***Row 2: Isonychia, Talleur's Golden Stone, Gray Nymph***
***Row 3: Hendrickson Poly-Fluff Emerger, Western Green Drake Emerger, Whitlock Fox Squirrel Nymph***
***Row 4: Balloon Emerger, PMD Emerger, Compara-Emerger*** Photograph by Tim Savard.

*Ribbing:* Fine copper wire or oval tinsel.

*Body/Thorax:* Light olive dubbing.

*Hackle (Legs):* Hungarian partridge barbs.

*Underwing:* Gray cul de canard, tied on in a clump.

*Overwing:* Gray Antron-type yarn.

*Comments:* This dressing successfully imitates the emerging forms of the several mayflies that comprise the western green drake group. The fly is commonly tied in sizes 8 to 12, on 1X- or 2X-long hooks.

## Streamers, et cetera

### *Dick's Gray Ghost*

*Body:* Pumpkin orange floss.

*Tag/Ribbing:* Narrow, flat silver tinsel.

*Belly:* A small bundle of peacock herl atop a small bunch of white bucktail.

*Underwing:* A long, straight golden-pheasant crest feather laid along the top of the body.

*Wing:* Gray streamer hackles.

*Cheek/Eye Assembly:* Silver pheasant saddles, Mylar sheet eyes.

*Throat (optional, omitted here):* A few short strands of brilliant yellow synthetic floss.

*Comments:* This dressing contains a few modifications of the Carrie Stevens original, which called for jungle cock eyes and a couple of short golden-pheasant crest feathers at the throat. Construct the cheek/eye assembly ahead of time thusly: First, stick the eye to the cheek feather. Then, with a toothpick, rub a small droplet of Zap-A-Gap or similar adhesive on the backside, but only where the eye is situated. It will soak through and bond the eyes. Let dry before tying on. Also, if you don't want to fool with golden-pheasant crest, substitute a small bunch of rainbow Krystal Flash, tied in the same position along the back. This fly is usually tied fairly large.

### *Olive Matuka*

*Body:* White or silver gray Nylon Stretch.

*Ribbing:* Oval silver tinsel.

*Wing:* Two dense, webby, olive-dyed grizzly body or hen saddle feathers, concave-to-concave; pluck the bottoms off the barbs to fit over the body.

*Collar:* More of the dyed grizzly, tied collar style, and full.

*Comments:* The technique here is the same as for the Rabbit Matuka—the body and ribbing materials are both mounted in bobbins. Feel free to substitute other types of front ends for the collar, such as a Muddler spun-and-trimmed-hair assembly in olive, dark brown, or black. And feel free to tie this fly in whatever other color appeals to you. It's commonly tied in sizes from 2 to 8.

### *Muddler Minnow*

*Tail:* Two sections of mottled turkey.

*Underbody:* Lead wire.

*Body:* Gold braid or mesh.

*Underwing:* A small bunch of gray- or fox-squirrel tail hair.

*Wing:* Same as tail.

*Collar/Head:* Spun and clipped deer hair.

*Comments:* Here, you can utilize the techniques you learned in the Rabbit Matuka exercise. Note that while the tail and wings are tied in the fashion of a standard wet fly, both pairs of slips are inverted, so that the tips point downward. It's important to completely secure the wings before moving on to the collar/head, so that the spinning activity doesn't upset them. Don't be discouraged if this fly doesn't come out just right the first time; it takes a bit of practice. The Muddler is tied in a wide range of sizes, from tiny to huge; 4 to 10 are the most common.

### *Marabou Muddler—White*

*Tail:* A bunch of red hackle barbs.

*Underbody:* Lead wire.

*Body:* Krystal Chenille; pearl, silver, or gold.

*Wing:* White marabou.

*Overwing:* A small bunch of peacock herl.

*Collar/Head:* Same as for the Muddler Minnow.

*Comments:* Marabou Muddlers are dynamite flies. They're commonly tied in yellow and black, as well as white. Sizes range from 6 to 2/0.

### *Marabou Black Ghost*

*Tail:* A bunch of yellow hackle barbs.

*Ribbing:* Narrow flat or oval silver tinsel.

*Body:* Black flosslike material.

*Throat:* Yellow hackle barbs.

*Wing:* White marabou.

*Eye (optional):* Painted, or Mylar sheet; epoxied.

*Comments:* Why this fly works so well is anyone's guess; the fact is that it does. The original dressing specified jungle cock eyes; however, my experience indicates that the type of eye shown here works even better. The fly is commonly tied in sizes 2 to 10.

### *Mickey Finn*

*Ribbing:* Fairly heavy oval silver tinsel.

*Body:* Fairly wide flat silver tinsel.

*Wing:* A small bunch of yellow hair, over which is a small bunch of red hair; over this is another bunch of yellow hair, equal in quantity to the first two bunches combined.

*Eye (optional):* Same as for the Black Ghost.

*Comments:* An all-time world-renowned streamer that has taken everything from Atlantic salmon to northern pike to saltwater fish. Feel free to substitute an embossed-tinsel body, without ribbing, for the one specified in the dressing. Be careful to tie the wing in layered fashion, as per the Black-Nosed Dace exercise. The original dressing did not call for an eye, but as you've undoubtedly figured out by now—I like eyed streamers. The fly is commonly tied from sizes 2 to 12.

### *Hornberg*

*Body:* Flat silver tinsel.

*Underwings:* Yellow hackles, tied flat against the sides of the hook.

*Overwings:* Mallard flank feathers, also tied flat against the sides of the hook.

*Cheeks:* Jungle cock eye or substitute.

*Hackle:* Soft grizzly, tied collar style.

*Comments:* Also called the Hornberg Special, this is a murderous fly everywhere, on many species of fish. The body and underwings aren't clearly visible, but when the fly is wet, they glow through.

Note what appear to be real jungle cock cheeks; they aren't! Just as I was completing this book, a new product hit the market: phony jungle cock on synthetic sheeting, made to look realistic via computergraphic technology. This particular brand is Featherite, a product of Western Flycraft, a Scottish company. Several sizes are available.

This is the first fly I've tied with the new material. I formed an assembly by Zap-A-Gapping the eyes to the mallard flanks ahead of time. I believe the use of an adhesive is a must, no matter what the wing is made of. This product, barring unforeseen problems, should solve the jungle cock dilemma for everything except collectible "wall-flies." Tie the Hornberg in a size range of 4 to 10.

### *Clouser Minnow—Sculpin*

*Eyes:* Dumbbell.

*Belly:* White hair.

*Glitz:* Bronze-gold Krystal Flash.

*Wing:* Fox-squirrel tail.

*Comments:* As I mentioned in the pattern lesson in chapter 5, the Clouser Minnow is one of the most versatile and adaptable styles of fly ever developed. Here, the components are chosen to simulate a favorite forage fish, the sculpin. The fly is tied from 1½ to 4 inches (37–100mm) in overall length.

### *Clouser Minnow—Baby Rainbow Trout*

*Eyes:* Silver bead chain or small dumbbell.

*Belly:* White hair.

*Wing Assembly:* In this order, working from bottom to top: pink hair, green hair, rainbow Krystal Flash, gray hair.

*Comments:* This version of the Clouser Minnow is intended to appeal to the appetite of large, voracious, and perhaps cannibalistic trout. It's tied from 2 to 4 inches (50–100mm) in overall length.

### *Edson Tiger—Light*

*Tag:* Narrow, flat gold tinsel; just a turn or two at the very bend of the hook, where it begins to slope.

*Tail (optional):* Black-and-white-tipped wood duck.

*Body:* Peacock herl.

*Wing:* Yellow hair.

*Neck Accent:* Two small red-dyed hen hackles.

*Cheeks:* Jungle cock or substitute.

*Comments:* As the name implies, there's also a dark Edson Tiger. Both are time-proven performers. Usually tied in sizes 4 to 8.

### *Green-Butt Skunk*

*Hook:* Salmon wet-fly.

*Tail:* A bunch of red-dyed hackle barbs.

*Butt:* Fluorescent green chenille or yarn.

*Ribbing:* Oval silver tinsel.

*Body:* Black chenille.

*Wing:* White hair: calf tail, goat, bucktail, or whatever.

*Collar:* Soft black hackle.

*Comments:* This is a famous steelhead fly. It's commonly tied in sizes 4 to 12.

### *Yellow Cosseboom*

*Thread:* Red.

*Hook:* Salmon wet-fly.

*Tail:* A short piece of yellow floss.

**Streamers, *Left to Right, From the Top:***
***Top: Marabou Muddler—White, Edson Tiger—Light***
***Center, Top: Muddler Minnow***
***Center, Bottom: Dick's Gray Ghost***
***Bottom: Clouser Minnow—Sculpin, Clouser Minnow—Baby Rainbow Trout***
Photograph by Tim Savard.

*Tag/Ribbing:* Narrow, embossed silver tinsel.

*Body:* Yellow floss.

*Wing:* Gray-squirrel tail.

*Collar:* Soft grizzly.

*Comments:* This fly is a longtime favorite of those who pursue *Salmo salar.* The tinsel is tied in

**Streamers, *Left to Right, From the Top:***
***Row 1: Green-Butt Skunk, Yellow Cosseboom***
***Row 2: Hornberg, Olive Matuka***
***Row 3: Marabou Black Ghost, Mickey Finn*** Photograph by Tim Savard.

after the tail is tied on. Then the tag and ribbing are applied in one continuous operation. The fly is commonly tied in sizes 4 to 10.

There; I've thrown in a steelhead fly and an Atlantic salmon fly just to show you that a mere *steelhead* or *salmon* designation shouldn't deter the fledgling tyer. Fly tying is—well, fly tying! When looking through pattern books, deciding what dressings I wanted to show here, I was astounded by the number of flies any journeyman tyer could easily dress with the skills attained from the foregoing exercises.

I sincerely hope that fly tying will enhance not only your angling career, but your very lifestyle as well. It certainly has mine.

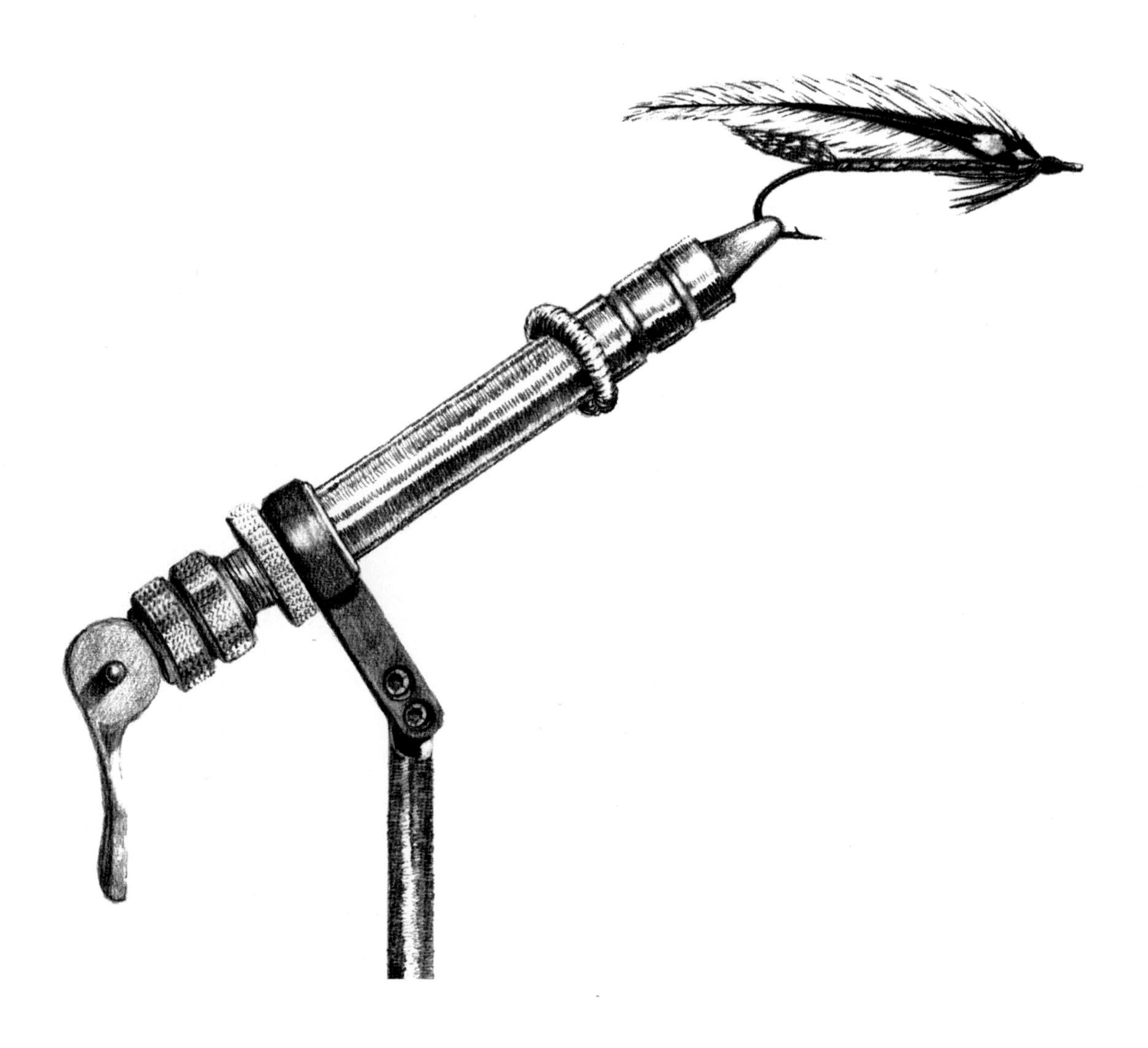

◂APPENDIX▸

# Hook Substitution Chart

I have a preference for Daiichi hooks. However, there are other quality brands available, and I'd be remiss if I didn't make mention of them and list like models.

The hooks listed here may vary a bit in size and design between manufacturers; however, they're similar enough that flies tied on them will come out virtually the same as those tied on the hooks specified in the exercises. In cases where the match is not exact but, in my judgment, the hook in question is close enough to cover comparable Daiichi models, I use a ditto mark. Here, I list only the models used in the tying lessons in the order in which they appear in the book. A complete and comprehensive substitution chart can be obtained from any Daiichi dealer.

**Hook Brands and Model Numbers**

| *Daiichi* | *Mustad* | *Orvis* | *Eagle Claw* | *Partridge* | *Tiemco* | *Belvoirdale* |
|---|---|---|---|---|---|---|
| 1720, 3XL | 9672 | J1526 | 58 | H1A | 5263 | Nymph hook |
| 2220, 4XL | 79580 | J8808 | 281 | D4A | " | " " |
| 1/10, 2XL | 9671 | J1524 | 63 | H1A | 5262 | " " |
| 1550, Std. | 3906 | J167T | 57 | G3A | 3761 | Wet Fly |
| 1560, 1XL | 3906B | " | " | " | " | " " |
| 2340, 6XL | 3665A | J1511 | ——— | ——— | 300 | Streamer |
| 2370, 7XL | 9575 | " | ——— | ——— | " | " |
| 1180, DF | 94840 | J1523 | 59 | E1A, L3A | 5210 | Dry Fly |
| 1100*, DF | 80000BR | J4864 | " | " | 100 | " " |

***Note*** *The Daiichi #1100 was not mentioned in the book. It's a dry-fly hook with a slightly wider than average gape.

# Index


Adams (fly pattern), 128, 130, 133*illus.*
Alder (fly pattern), 135, 137*illus.*

badger hackle, 82, 84
Balloon Emerger (fly pattern), 139-140, 141*illus.*
barbs, 19, 64, 91
barbules, 61
beard, 34, 45
biots, 57, 58
Black-Nosed Dace Bucktail (fly pattern), 74-75
Black Wulff (fly pattern), 105
Blue-Winged Olive (fly pattern), 102-104
bobbins, 6, 7*illus.*, 8-9
bobbin threaders/reamers, 7*illus.*, 9, 11
Brown Bivisible (fly pattern), 131, 133*illus.*
Brown-Hackle Peacock (fly pattern), 136, 137*illus.*
bucktail, 73-74
Bugskin, 55, 57*illus.*

Callibaetis Sparkledun (fly pattern), 132
cape hackle, 20, 91, 92-93
Catch-A-Hatch, 62
C-clamps, 1-3, 2*illus.*, 6
cement, 14, 49
chenille, 17, 20, 23*illus.*
Clouser Deep Minnow (fly pattern), 82
Clouser Minnow (fly pattern), 78-82, 144-145, 146*illus.*
collar hackle, 34
collets, 3, 5
combs, 11*illus.*, 13
comparadun wings, 116
Compara-Emerger (fly pattern), 140, 141*illus.*
Cow Dung (fly pattern), 134, 137*illus.*
Cream Variant (fly pattern), 99-101
cul de canard (CDC), 61

Dark Cahill Wet Fly (fly pattern), 43-45
Dick's Dace (fly pattern), 76-78
Dick's Gray Ghost (fly pattern), 142, 146*illus.*
Dorato Hare's Ear (fly pattern), 131-132, 133*illus.*
dry flies, 91-124, 126-128, 129*illus.*, 130-132, 133*illus.*
dubbing, 9, 12-13, 33, 36
Dun Variant (fly pattern), 98, 101, 127, 129*illus.*
dyed hackle, 46

Eastern March Brown (fly pattern), 126
Edison Tiger (fly pattern), 145, 146*illus.*
electronics clips, 58
Ellis' Fly (fly pattern), 136, 137*illus.*
emergers, 60-62
eyes, 77-79

Figure-8 Wraps, 106, 109*illus.*
Fish Fuzz, 80
flank feathers, 34, 45, 111-112
Flick, Art, 73, 75, 98, 99, 101
floss, 33, 46, 48*illus.*
Fly Fur, 80
fly patterns. *See* dry flies; *names of individual patterns;* nymphs; streamers; wet flies
fly tying, 10*illus.*
  bad habits of, 8
  budget for, 5
  dry flies, 91-124, 126-128, 129*illus.*, 130-132, 133*illus.*
  materials, 17-20
  nymphs, 51-62
  streamers, 63-89
  terms, 21, 30, 33-34, 38, 42, 51, 55, 58, 65-66, 75-76, 91-92, 99, 101, 106
  vises, 1-6
  vision problems and, 15
  wet flies, 33-49, 134-142, 137*illus.*, 141*illus.*

Girdle Bug (fly pattern), 28, 29-30
Gold Flashback Nymph (fly pattern), 52-54
Gold-Ribbed Hare's Ear (fly pattern), 135, 137*illus.*
Grannom (fly pattern), 134, 137*illus.*
Gray Fox Variant (fly pattern), 98, 101, 127-128, 129*illus.*
Gray Nymph (fly pattern), 138, 141*illus.*
Green-Butt Skunk (fly pattern), 145, 147*illus.*
Griffin 1-A vise, 2*illus.*
Griffin bobbin, 7*illus.*
Griffith Gnat (fly pattern), 132, 133*illus.*
Grizzly Bivisible (fly pattern), 96-98
Grizzly-Hackle Peacock (fly pattern), 136, 137*illus.*
Grizzly King (fly pattern), 135, 137*illus.*
Grizzly Wulff (fly pattern), 127, 129*illus.*
Grouse & Green Soft-Hackle (fly pattern), 136, 137*illus.*

hackle, 20, 21*illus.*, 25*illus.*, 34, 46, 92, 93
hackle gauges, 12
hackle guards, 11*illus.*, 13
hackle pliers, 7*illus.*, 9, 24
hair eveners, 11*illus.*, 11-12, 105
hair-packing tools, 11*illus.*
Hair-Wing Caddis (fly pattern), 119-120, 133*illus.*
Hare's Ear Soft-Hackle (fly pattern), 35-38
Hatches (Caucci and Nastasi), 116
head lacquer, 13, 14, 27, 49
Hemingway Caddis (fly pattern), 130-131, 133*illus.*
Hendrickson Poly-Fluff Emerger (fly pattern), 140, 141*illus.*
Henryville Special (fly pattern), 130, 133*illus.*
hooks, 3-5, 17-19
  for dry flies, 93, 96
  mounting, 6
  for streamers, 63-65
  substitution chart, 149*t*
  for wet flies, 34-35
Hornberg (fly pattern), 144, 147*illus.*
House & Lot (fly pattern), 127, 129*illus.*

Hungarian partridge feathers, 34
Hunter's Multi-Head vise, 4

integration (tying procedure), 84
iris scissors, 7*illus.*
Iron Blue Dun (fly pattern), 39–42
Irresistible (fly pattern), 84
isolation (tying procedure), 84
Isonychia (fly pattern), 139, 141*illus.*

jungle cock, 66

Krystal Flash, 79

lacquer, 13, 14, 49
lamps, 14–15
Leadwing Coachman (fly pattern), 134, 137*illus.*
Light Cahill (fly pattern), 130, 133*illus.*
lighting, 14–15
Light (Pink) Hendrickson (fly pattern), 128, 129*illus.*
Light Spruce Fly (fly pattern), 82–84
Lime Trude (fly pattern), 129*illus.*, 131

marabou, 25–26, 27*illus.*, 71–73
Marabou Black Ghost (fly pattern), 143, 147*illus.*
Marabou Muddler (fly pattern), 143, 146*illus.*
Matarelli bobbin, 7*illus.*
Matarelli whip-finishing tool, 7*illus.*, 111
materials clips, 12
Mickey Finn (fly pattern), 73, 143–144, 147*illus.*
*Modern Fly-Tying Materials* (Talleur), 1, 20
Montana Nymph (fly pattern), 30-32
Montreal (fly pattern), 135, 137*illus.*
Muddler Minnow (fly pattern), 8, 21, 84, 142–143, 146*illus.*

nymphs, 18, 26, 51–62, 136–140, 141*illus.*, 142

Olive CDC Emerger (fly pattern), 61-62
Olive Drake Nymph (fly pattern), 139, 141*illus.*
Olive Grizzly Woolly Worm (fly pattern), 22–25
Olive Matuka (fly pattern), 142, 147*illus.*
Olive Sparkledun (fly pattern), 133*illus.*

Pale Morning Dun (fly pattern), 131, 133*illus.*
palmered hackle, 21, 96
peacock herl, 57–58, 82, 112–113
pheasant tail, 55
Pheasant Tail (P-T) Nymph (fly pattern), 55–57
Picket Pin (fly pattern), 134
pin vises, 7*illus.*, 13
PMD Emerger (fly pattern), 140, 141*illus.*
Prince Nymph (fly pattern), 57–60
Professor, The (fly pattern), 47–49
Profile Plate, 15

Quill Gordon (fly pattern), 128

Rabbit Matuka (fly pattern), 85–88
Rat-Faced MacDougal (fly pattern), 84
Red Quill (fly pattern), 128, 129*illus.*
Regal vise, 3–4, 4*illus.*, 6
return wire (hooks), 64
reverse-ribbing, 55
ribbing, 34, 46
rotary vises, 5–6
Royal Coachman, 105
Royal Wulff (fly pattern), 126, 129*illus.*
rubber legs, 28
Rusty Spinner (fly pattern), 121–123

saddle hackle, 20, 23*illus.*
San Juan Worm (fly pattern), 20
scissors, 7*illus.*, 7–8
Shushan Postmaster (fly pattern), 66–73
Sierra Red Dot (fly pattern), 131, 133*illus.*
spade hackle, 92
spent-wings, 120–121
squirrel tail, 66
stackers, 11–12
stacking, 85, 105
stationary vises, 1, 6
SteppenWulff (fly pattern), 106–111
streamers, 18, 25, 63–89, 142–147, 146–147*illus.*
Streamside Guide to Naturals and Their Imitations, A (Flick), 98
stripped quill, 99
strung herl, 58
sweet spot, 91
synthetic hair, 79–80

tails, 113, 124
Talleur's Golden Stone (fly pattern), 138, 141*illus.*
Tan/Grizzly Caddis (fly pattern), 130
Teal-Winged Adams Quill (fly pattern), 113–115
Tellico (fly pattern), 138, 141*illus.*
Thompson Pro vise, 2*illus.*
threads, 19–20
thread torque, 21, 125
throat hackle, 92
tinsel, 34
tippet wings, 101
tools, 7–9, 10*illus.*, 11–15
toothpicks, 13
Troth Caddis (fly pattern), 118
true hackle, 34, 45, 65–66
tubing, 9, 75
tweezers, 11*illus.*, 12, 28

Ultra-chenille, 20
Umpqua bobbin, 7*illus.*
Uni-Thread, 20, 46, 85

Variant flies, 98, 99
Vernille, 20
vises, 1, 2*illus.*, 2–6
vision problems, 15
Voyager vise, 4*illus.*

Western Green Drake Emerger (fly pattern), 140–142, 141*illus.*
Western Green Drake (fly pattern), 129*illus.*, 132
Western Pale Morning Dun Sparkledun (fly pattern), 117–118
wet flies, 33–39, 134–142, 137*illus.*, 141*illus.*
whip-finish tools, 7*illus.*, 11
White Wulff (fly pattern), 127, 129*illus.*
Whitlock Fox Squirrel Nymph (fly pattern), 141*illus.*
Whitlock Red Squirrel Nymph (fly pattern), 138–139
wing burners, 11*illus.*, 13–14
wing cases, 30
wing quills, 34, 38–39
Woolly Bugger (fly pattern), 25, 26–28, 79
Woolly Worm (fly pattern), 17, 20

X-Wraps, 106, 108*illus.*

Yellow Cosseboom (fly pattern), 145–147, 147*illus.*

Zap-A-Gap, 13, 27, 80–81, 99
zonker strips, 89
Zug Bug (fly pattern), 136–137, 141*illus.*